BASICS OF DIGITAL ELECTRONICS

BASICS OF DIGITAL ELECTRONICS

BANANI GHOSH

B. Tech (Electronics and Communication Engineering)
Birla Institute of Technology,
Mesra, Ranchi

CRC Press
Taylor & Francis Group
Boca Raton London New York

CRC Press is an imprint of the
Taylor & Francis Group, an **informa** business

Manakin
PRESS

First published 2024
by CRC Press
4 Park Square, Milton Park, Abingdon, Oxon, OX14 4RN

and by CRC Press
2385 NW Executive Center Drive, Suite 320, Boca Raton FL 33431

CRC Press is an imprint of Informa UK Limited

© 2024 Manakin Press

The right of Banani Ghosh to be identified as author of this work has been asserted in accordance with sections 77 and 78 of the Copyright, Designs and Patents Act 1988.

Print edition not for sale in South Asia (India, Sri Lanka, Nepal, Bangladesh, Pakistan or Bhutan)

ISBN: 9781032567556 (hbk)
ISBN: 9781032567563 (pbk)
ISBN: 9781003436997 (ebk)

DOI: 10.4324/9781003436997

Typeset in Times New Roman
by Manakin Press, Delhi

Manakin
PRESS

Brief Contents

Detailed Contents

1

Analog and Digital Signals

FACTS THAT MATTER

Signal:	A signal is the transmission of *data* that we constantly deal with during our daily routine life.
Analog signal:	A continuous signal whose amplitude can take any value between given limits.
Digital signal:	A signal that changes amplitude in discrete steps and thus its amplitude can have only discrete values between given limits.
Bit:	Binary digit.
Binary number:	A number code that uses only the digits 0 and 1 to represent quantities.
Clock:	A periodic rectangular waveform used as a basic timing signal.
Flip-flop:	An electronic circuit that can store one bit of a binary number.
Memory element:	Any device or circuit used to store 1 bit of a binary number. Thus flip-flop is a memory element.
A/D converter:	Converts an analog signal to digital signal.
D/A converter:	Converts digital signal to an analog signal.
Hybrid system:	Such systems in which combination of analog and digital systems are used.

OBJECTIVE TYPE QUESTIONS

Fill in the Blanks:

1. Analog signal is a _____ signal.
2. Digital signal has _____ levels.
3. The operation of a digital circuit is generally considered to be _____.
4. A binary digit is known as _____.
5. Such systems in which combination of analog and digital systems are used are known as _____.

Answer

1. continuous.
2. two discrete.
3. nonlinear.
4. bit.
5. hybrid systems.

LONG ANSWER TYPE QUESTIONS

Q.1. Differentiate between analog signal and digital signal.

Ans. Differences between analog signal and digital signal are as follows:

Analog signal	Digital signal
1. Analog signal is a continuous electrical signal.	1. Digital signal is a discrete electrical signal.
2. It is in the form of a sinusoidal waveform.	2. It is in the form of square wave or pulses.

3. It performs linear operations.	3. It performs nonlinear operations.
4. It is used to represent analog data.	4. It is used to represent digital data.
5. Analog data is a function of time and occupy a limited frequency spectrum which can be represented by an electromagnetic signal.	5. Digital data can be represented by digital signals with a different voltage level for each of the two binary digits 0 or 1.
6. It is less reliable as it can have any value between the given range	6. It is more reliable and simple as it has only two states.
7. It cannot provide noise free operation.	7. It provides noise free operation.
8. Difficult to integrate different source of information using analog signal.	8. Easy to integrate different source of information using digital signals.

(Contd...)

Analog signal	Digital signal
9. Analog display is less accurate than digital display as human error may be there.	9. Digital display is very accurate as human error is eliminated.
10. High degree of security of information cannot be maintained in analog transmission.	10. High degree of security of information can be maintained in digital transmission.
11. Analog signals do not have the potential to be stored, retrieved, processed and manipulated for signal enhancement as these signals first travel into the modem where they are transformed to digital signals for computer processing and then eventually modulate the signals.	11. Digital signals which are inherently compatible with computers have a potential to be stored, retrieved, processed and manipulated for signal enhancement and improved performance.
12. Analog signals can be processed by analog computers but there can be errors because of signals loss, interference and noise.	12. Digital signals consisting of only two values 0 (off or low) or 1 (on or high), are much less susceptible to transmission problems.
13. The channel bandwidth required for analog communication is less than digital communication.	13. The digital communication requires higher channel bandwidth than analog communication.

Q.2. What are the advantages of using digital signals over analog signals?

Ans. The main advantages of using digital signals over analog signals are as follows:

1. Digital signals have only two values LOW or 0 and HIGH or 1 and hence very simple and accurate as compared to analog signal which can have any value in a given range.

2. Digital signals can be processed and transmitted more efficiently and reliably than analog signals.

3. Digital signals provide grater immunity to noise when compared with the analog signals.

4. There are various means of detecting and correcting the errors occurred during the process of transmission of digital signals which is not possible for transmission of analog signals.

5. Digital signals, which are inherently compatible with computers, have a potential to be stored, retrieved, processed and manipulated for signal enhancement and improved performance. This is not possible for analog signals.

6. Using digital signals, it is easy to integrate different sources of information into a common format which is difficult using analog signals.

7. High degree of security of information can be maintained in the course of transmission using digital signals unlike transmission using analog signals.

8. The digital display is more easier to read than the analog display.

9. The size of digital device using digital signals is very small due to digital ICs and these ICs are very cheap compared to the size of an analog device using analog signal which is more costlier also.

10. Digital communication is readily adaptive to other powerful and advanced data processing such as digital signal processing, image processing etc. This is not possible with analog communication.

11. System with higher complexities can be built in a cost effective manner using more reliable and specific ICs compared to analog systems.

12. Digital system provides more flexibility in configuring digital communication systems than analog system.

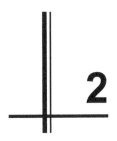

Number Systems

FACTS THAT MATTER

Number system:	A number system is a way to represent numbers.
Base or radix:	The number of digits or basic symbols used in a number system is known as base or radix.
Decimal system:	Refers to a number system with a base of 10. These are 0 through 9.
Binary system:	Refers to a number system with a base of 2. These are 0 and 1.
Octal system:	Refers to a number system with a base of 8. These are 0 through 7.
Heradecimal system:	Refers to a number system with a base of 16. It has digits 0 through 9 followed by A through F.
Bit:	An abbreviated form of binary digit. For *e.g.* 11011 has 5 binary digits, we can say that it has 5 bits.
Byte:	A binary number with 8 bits is called a byte.
Nibble:	A binary number with 4 bits is called nibble.
Double Dabble:	It is a method for converting decimal numbers to binary numbers.
Fixed point representation:	In fixed point representation, the location of the decimal position is fixed in the number string. For *e.g.* in 8 bit representation if the decimal position is fixed in

such a way that 5 bits are available for integer, then 8–5 = 3 bits are available for fractional part. This is often written as fixed (8, 3) representation.

Floating point representation:

In floating point representation, the decimal point position floats *i.e.*, not fixed. The number representation has two parts—mantissa (M) and exponent (E). The number for base 2 is represented as $M \times 2^E$. The decimal point depends on value of E.

Weight:

Refers to the decimal value of each digit position of a number.

e.g. in decimal number system

$$\leftarrow 10^4 \; 10^3 \; 10^2 \; 10^1 \; 10^0 \cdot 10^{-1} \; 10^{-2} \; 10^{-3} \rightarrow$$
$$\uparrow$$
decimal point

∴ the weights are 1, 10, 100, 1000 ... and so on working from the decimal point to the left and $\dfrac{1}{10}, \dfrac{1}{100}, \dfrac{1}{1000} \cdots$ and so on working from the decimal point to the right.

In binary number system

$$\leftarrow 2^3 \; 2^2 \; 2^1 \; 2^0 \cdot 2^{-1} \; 2^{-2} \; 2^{-3} \rightarrow$$
$$\uparrow$$
binary point

∴ the weights are 1, 2, 4, 8 ... and so on starting from the binary point to the left and $\dfrac{1}{2}, \dfrac{1}{4}, \dfrac{1}{8} \cdots$ and so on starting from the binary point to the right.

In octal number system

$$\leftarrow 8^3 \; 8^2 \; 8^1 \; 8^0 \cdot 8^{-1} \; 8^{-2} \; 8^{-3} \rightarrow$$
$$\uparrow$$
octal point

∴ the weights are 1, 8, 64 ... and so on starting from the octal point to the left and $\dfrac{1}{8}, \dfrac{1}{64}, \dfrac{1}{512} \cdots$ and so on starting from the octal point to the right.

In hexadecimal number system

$$\leftarrow 16^3 \; 16^2 \; 16^1 \; 16^0 \cdot 16^{-1} \; 16^{-2} \; 16^{-3} \rightarrow$$
$$\uparrow$$
hexadecimal point

∴ the weights are 1, 16, 256 ... and so on starting from the hexadecimal point to the left and $\dfrac{1}{16}, \dfrac{1}{256} \cdots$ and so on starting from the hexadecimal point to the right.

Conversion:

Decimal to Binary: To convert a decimal number into binary, we divide the number successively by 2. The remainder when read upward gives the binary number. In case of fraction the number is multiplied by 2 successively. The carry when read downward gives binary number.

Binary to decimal: To convert a binary number in the decimal number we multiply each binary digit by its weight. The weights are $2^0, 2^1, 2^2, 2^3$ ... starting with the least significant bit and take the sum which gives the decimal number. *For fractional part* we multiply each bit by its weight. The weights are $2^{-1}, 2^{-2}, 2^{-3}$ ... starting from left to right after the binary point and take the sum which gives the decimal number.

Decimal to octal: To convert a decimal number in to octal, we divide the number successively by 8. The remainder when read upward gives the octal number. In case of fraction the number is multiplied by 8 successively. The carry when read downward gives octal number.

Octal to decimal: To convert an octal number to decimal number, we multiply each octal digit by its weight. The weights are $8^0, 8^1, 8^2$ ... starting with its least significant digit and take the sum which gives the decimal number. *For fractional part* we multiply each digit by its weight. The weights are $8^{-1}, 8^{-2}, 8^{-3}$ ... starting from left to right after the octal point and take the sum which gives the decimal number.

Decimal to hexadecimal: To convert a decimal number into hexadecimal we divide the number successively by 16. The remainder when read upward gives the hexadecimal number. In case of fraction the number is multiplied by 16 successively. The carry when read downward gives hexadecimal number.

Hexadecimal to decimal: To convert a hexadecimal number to decimal number, we multiply each hexadecimal digit by its weight. The weights are 16^0, 16^1, 16^2 ... starting with least significant digit and take the sum which gives the decimal number. *For fractional part* we multiply each digit by its weight. The weights are 16^{-1}, 16^{-2}, 16^{-3} ... starting from left to right after the hexadecimal point and take the sum which gives the decimal number.

OBJECTIVE TYPE QUESTIONS

1. How many bits are required to represent decimal 15?
2. What is the binary equivalent of decimal 12.
3. What i the decimal equivalent of binary 1001?
4. A binary number has 9 bits. What is the binary weight of the most significant bit?
5. What is the main difference between fixed point and floating point representation?
6. How many bits are available after decimal point in fixed (8, 5) representation?
7. Which of fixed (8, 5) and fixed (8, 3) has higher precision?
8. What is double dabble?
9. Differentiate between binary, hexadecimal and decimal number system.
10. Why is hex more efficient than binary?
11. How many bits are there in 2 bytes?
12. How many nibbles are there in (a) 11110000 (b) 110010101111?

Answers

1. Four.
2. 1100.
3. 9.
4. $2^8 = 256$.
5. Position of decimal point is fixed in fixed-point representation but floats or not fixed in floating-point representation.
6. 5 bits.
7. Fixed (8, 5) has higher precision.
8. Double-dabble is a method for converting decimal numbers to binary numbers.
9. The difference between the three number systems is that binary uses only two digits or bits 0 and 1, decimal number system uses 10 digits, 0 through 9 and hexadecimal number system uses 16 digits. 0 through 9 followed by A through F.

10. Hex or (hexadecimal) is a base 16 system used to simplify how binary is represented. An 8 bit binary number can be written using only two hex digits one hex for each nibble. It is much easier to write numbers as hex than to write them as binary.

11. 16 bits.

12. (*a*) 2, (*b*) 3.

SHORT ANSWER TYPE QUESTIONS

Q.1. Convert the following decimal numbers to binary numbers.

(*i*) 65, (*ii*) 10.7, (*iii*) 26.25, (*iv*) 63.362

Ans. (*i*) Using double dabble method

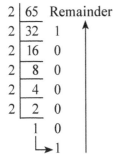

∴ Now taking remainder from bottom to top gives the equivalent binary.

∴ $65_{10} = 1000001_2$ **Ans.**

(*ii*) 10.7_{10}

Let us first convert the integer part 10_{10}

```
2 | 10   Remainder
2 |  5    0
2 |  2    1
2 |  1    0
     └► 1
```

Now taking remainder from bottom to top gives the equivalent binary.

∴ $10_{10} = 1010_2$

Next conversion of fractional part 0.7_{10}

$0.7 \times 2 = 1.4 = 0.4$ with a carry of 1

$0.4 \times 2 = 0.8 = 0.8$ with a carry of 0

$0.8 \times 2 = 1.6 = 0.6$ with a carry of 1

$0.6 \times 2 = 1.2 = 0.2$ with a carry of 1

Taking carry from top to bottom gives the equivalent binary

∴ $0.7_{10} \simeq 0.1011_2$

∴ $10.7_{10} \simeq 1010.1011_2$ **Ans**

(*iii*) 26.25_{10}

Let s first convert the integer part 26_{10}

```
2 | 26   Remainder
2 | 13    0  ↑
2 |  6    1
2    3    0
     1    1
     └→ 1
```

Taking remainder from bottom to top gives the equivalent binary

∴ $26_{10} = 11010_2$

Next conversion of fractional part 0.25_{10}

$0.25 \times 2 = 0.50 = 0.5$ with a carry of 0

$0.5 \times 2 = 1.0 = 0.0$ with a carry of 1 ↓

Taking carry from top to bottom gives the equivalent binary

∴ $0.25_{10} = 0.01_2$

∴ $26.25_{10} = 11010.01_2$ **Ans**

(*iv*) 63.362_{10}

Let s first convert the integer part 63_{10}

```
2 | 63   Remainder
2 | 31    1  ↑
2 | 15    1
2 |  7    1
2 |  3    1
     1    1
     └→ 1
```

Taking remainder from bottom to top gives the equivalent binary

∴ $63_{10} = 111111_2$

Next conversion of fractional part 0.362_{10}

$0.362 \times 2 = 0.724 = 0.724$ with a carry of 0 ↑

$0.724 \times 2 = 1.448 = 0.448$ with a carry of 1

$0.448 \times 2 = 0.896 = 0.896$ with a carry of 0

$0.896 \times 2 = 1.792 = 0.792$ with a carry of 1

$0.792 \times 2 = 1.584 = 0.584$ with a carry of 1

Taking carry from top to bottom gives the equivalent binary

$\therefore\ 0.362_{10} \simeq 0.01011_2$

$\therefore\ 63.362_{10} \simeq 111111.01011_2$ **Ans**

Q.2. Convert the following decimal numbers to octal numbers

(i) 13.82, (ii) 4096

Ans. (i) Let us first take the integer part 13_{10}

$$8\ \underline{|\ 13}\ \ \text{Remainder}$$
$$1\quad 5 \uparrow$$
$$\hookrightarrow 1$$

Taking remainder from bottom to top gives the equivalent octal

$\therefore\ 13_{10} = 15_8$

Next conversion of fractional part 0.82_{10}

$0.82 \times 8 = 6.56 = 0.56$ with a carry of 6

$0.56 \times 8 = 4.48 = 0.48$ with a carry of 4

$0.48 \times 8 = 3.84 = 0.84$ with a carry of 3 ↓

Taking the carry from top to bottom gives the equivalent octal

$\therefore\ 0.82_{10} \simeq 0.643_8$

$\therefore\ 13.82_{10} \simeq 15.643_8$ **Ans.**

(ii) 4096_{10}

Using octal dabble method

$$
\begin{array}{r|r|l}
8 & 4096 & \text{Remainder} \\
8 & 512 & 0 \uparrow \\
8 & 64 & 0 \\
8 & 8 & 0 \\
 & 1 & 0 \\
 & & \hookrightarrow 1
\end{array}
$$

Taking remainder from bottom to top gives the equivalent octal

$\therefore\ 4096_{10} = 10000_8$ **Ans.**

Q.3. Convert the following decimal numbers to hexadecimal numbers

(i) 4096, (ii) 28.8125

Ans. (i) Using hex dabble method

```
16 | 4096   Remainder
16 |  256   0
16 |   16   0
       1    0
       └→1
```

Taking remainder from bottom to top gives the equivalent hexadecimal

$\therefore 4096_{10} = 1000_{16}$ **Ans.**

(ii) 28.8125

Let us first convert the integer part 28_{10}

```
16 | 28   Remainder
     1  C
     └→1
```

Taking remainder from bottom to top gives the equivalent hexadecimal

$\therefore 28_{10} = 1C_{16}$

Next conversion of fractional part 0.8125_{10}

$0.8125 \times 16 = 13.00 = 0.0$ with a carry of 13 i.e. D

$\therefore 28.8125_{10} = 1C.D_{16}$ **Ans.**

Q.4. Convert the following binary numbers to decimal numbers

(i) 10111100, (ii) 11001.011 (iii) 11010.111

Ans. (i) $10111100_2 = 1 \times 2^7 + 0 \times 2^6 + 1 \times 2^5 + 1 \times 2^4 + 1 \times 2^3 + 1 \times 2^2$
$$+ 0 \times 2^1 + 0 \times 2^0$$
$$= 128 + 0 + 32 + 16 + 8 + 4 + 0 + 0$$
$$= 188_{10} \quad \textbf{Ans.}$$

(ii) $11001.011_2 = 1 \times 2^4 + 1 \times 2^3 + 0 \times 2^2 + 0 \times 2^1 + 1 \times 2^0 + 0 \times 2^{-1}$
$$+ 1 \times 2^{-2} + 1 \times 2^{-3}$$
$$= 16 + 8 + 0 + 0 + 1 + 0 + 0.25 + 0.125$$
$$= 25.375_{10} \quad \textbf{Ans.}$$

(iii) $11010.111_2 = 1 \times 2^4 + 1 \times 2^3 + 0 \times 2^2 + 1 \times 2^1 + 0 \times 2^0 + 1 \times 2^{-1}$
$$+ 1 \times 2^{-2} + 1 \times 2^{-3}$$
$$= 16 + 8 + 0 + 2 + 0 + 0.5 + 0.25 + 0.125$$
$$= 26.875_{10} \quad \textbf{Ans.}$$

Q.5. Convert the following octal numbers to decimal numbers

(i) 216, (ii) 736.5

Ans. (i) $216_8 = 2 \times 8^2 + 1 \times 8^1 + 6 \times 8^0$
$$= 128 + 8 + 6 = 142_{10} \quad \textbf{Ans.}$$

(ii) $736.5_8 = 7 \times 8^2 + 3 \times 8^1 + 6 \times 8^0 + 5 \times 8^{-1}$

$= 448 + 24 + 6 + 5 \times \dfrac{1}{8}$

$= 478.63_{10}$ **Ans.**

Q.6. Convert the following hexadecimal numbers to decimal numbers

(i) 5A9, (ii) 3FA.8

Ans. (i) $5A9_{16} = 5 \times 16^2 + 10 \times 16^1 + 9 \times 16^0$

$= 1280 + 160 + 9$

$= 1449_{10}$ **Ans.**

(ii) $3FA.8_{16} = 3 \times 16^2 + 15 \times 16^1 + 10 \times 16^0 + 8 \times \dfrac{1}{16}$

$= 768 + 240 + 10 + 0.5$

$= 1018.5_{10}$ **Ans.**

Q.7. Convert the following octal numbers to binary numbers

(i) 567, (ii) 736.5.

Ans. (i) Here each of the given octal digit is converted to its equivalent 3-bit

$\therefore$ 5   6   7

↓   ↓   ↓

101   110   111

$\therefore$ $567_8 = 101110111_2$ **Ans.**

(ii) 736.5_8

Here each of the octal digits is converted to its equivalent 3-bit

$\therefore$ 7   3   6 · 5

↓   ↓   ↓   ↓

111   011   110 · 101

$\therefore$ $736.5_8 = 111011110.101_2$ **Ans.**

Q.8. Convert the following hexadecimal numbers to binary numbers

(i) C 5 E 2, (ii) B 4 D.15

Ans. (i) Here each of the given hexadecimal digit is converted to its equivalent
4-bit

$\therefore$ C   5   E   2

↓   ↓   ↓   ↓

1100   0101   1110   0010

$\therefore$ C 5 E $2_{16} = 1100010111100010_2$ **Ans.**

(ii) Here each of the given hexadecimal digits is converted to its equivalent
4-bit

$$\therefore \quad B \quad 4 \quad D \quad \cdot \quad 1 \quad 5$$
$$\downarrow \quad \downarrow \quad \downarrow \quad \quad \downarrow \quad \downarrow$$

1011 0100 1101 · 0001 0101

$\therefore$ B4D.15$_{16}$ = 101101001101.00010101$_2$ **Ans.**

Q.9. Convert the following binary numbers to octal

(*i*) 10101111, (*ii*) 1010011.1011011

Ans. (*i*) Converting the bits in groups of three, we get

010   101   111
$\downarrow$ $\downarrow$ $\downarrow$
2   5   7

$\therefore$ 10101111$_2$ = 257$_8$ **Ans.**

(*ii*) 1010011·1011011$_2$

Converting the bits in groups of three, we get

001 010 011 · 101 101 100
$\downarrow$ $\downarrow$ $\downarrow$ $\downarrow$ $\downarrow$ $\downarrow$
1   2   3 · 5   5   4

$\therefore$ 1010011.1011011$_2$ = 123.554$_8$ **Ans.**

Q.10. Convert the following binary numbers to hexa decimal numbers

(*i*) 10001100, (*ii*) 1010011.101101

Ans. (*i*) Converting the bits in groups of four, we get

1000   1100
$\downarrow$ $\downarrow$
8   C

$\therefore$ 10001100$_2$ = 8C$_{16}$

(*ii*) 1010011·101101$_2$

Converting the bits in groups of four, we get

0101   0011 · 1011   0100
$\downarrow$ $\downarrow$ $\downarrow$ $\downarrow$
5   3 ·   B   4

$\therefore$ 1010011.101101$_2$ = 53.B4$_{16}$ **Ans.**

Q.11. Convert the following hexadecimal numbers to octal numbers by first converting to binary

(*i*) 7AF4, (*ii*) B4D.E5

Ans. (*i*) To convert the hexadecimal number to binary, each of the given hexadecimal digits is converted *t* its equivalent 4 bit

$\therefore$ 7   A   F   4
    ↓   ↓   ↓   ↓

0111 1010 1111 0100

$\therefore$ $7AF4_{16} = 1111010111110100_2$

Now to convert from binary to octal, we have to convert the bits in groups of three

$\therefore$ 111   101   011   110   100
    ↓   ↓   ↓   ↓   ↓
    7   5   3   6   4

$\therefore$ $7AF4_{16} = 1111010111110100_2 = 75364_8$ **Ans.**

(*ii*) B4D.E5

To convert the hexadecimal number to binary, each of the given hexadecimal digits is converted to its equivalent 4 bit

$\therefore$ B   4   D · E   5
  ↓   ↓   ↓   ↓   ↓

1011 0100 1101 · 1110 0101

$\therefore$ $B4D.E5_{16} = 101101001101.11100101_2$

Now to convert from binary to octal, we have to convert the bits in groups of three

101   101   001   101 · 111   001   010
 ↓   ↓   ↓   ↓   ↓   ↓   ↓
 5   5   1   5 · 7   1   2

$\therefore$ $B4D.E5_{16} = 101101001101.11100101_2 = 5515.712_8$ **Ans.**

Q.12. Convert the following octal numbers to hexadecimal numbers by first converting to binary

(*i*) 325.736, (*ii*) 4073

Ans. (*i*) 325.736_8

To convert octal numbers to binary, each of the given octal digits is converted to its equivalent 3 bit

$\therefore$ 3   2   5 · 7   3   6
  ↓   ↓   ↓   ↓   ↓   ↓

011 010 101 · 111 011 110

$\therefore$ $325.736_8 = 11010101.11101111_2$

Now to convert from binary to hexadecimal numbers, we have to convert the bits in groups of four

$\therefore$ 1101 0101 · 1110 1111
   ↓   ↓   ↓   ↓
   D   5 · E   F

$\therefore$ $325.736_8 = 11010101.11101111_2 = D5.EF_{16}$ **Ans.**

(*ii*) 4073_8

To convert the octal number to binary, each of the given octal digits is converted to its equivalent 3 bit

$\therefore$ 4   0   7   3
       ↓   ↓   ↓   ↓
      100   000   111   011

$\therefore$ $4073_8 = 100000111011_2$

Now to convert from binary to hexadecimal numbers, we have to convert the bits in groups of four

$\therefore$ 1000   0011   1011
        ↓   ↓   ↓
        8   3   B

$\therefore$ $4073_8 = 100000111011_2 = 83B_{16}$ **Ans.**

Q.13. What are the hexadecimal number that follow each of these

(*i*) ABCD, (*ii*) 7F3F, (*iii*) BEEF

Ans. (*i*) $ABCD_{16}$ = 1010 1011 1100 1101
                                              +1
                          1010 1011 1100 1110
                            ↓   ↓   ↓   ↓
                            A   B   C   E

$\therefore$ ABCE follows ABCD   **Ans**

(*ii*) $7F3F_{16}$ = 0111 1111 0011 1111
                                       +1
                   0111 1111 0100 0000
                     ↓   ↓   ↓   ↓
                     7   F   4   0

$\therefore$ 7F40 follows 7F3F

(*iii*) $BEEF_{16}$ = 1011 1110 1110 1111
                                       +1
                   1011 1110 1111 0000
                     ↓   ↓   ↓   ↓
                     B   E   F   0

$\therefore$ BEF0 follows BEEF   **Ans**

Q.14. Find the decimal equivalent of 10110111 in fixed (8, 3) representation.

Ans. 10110111

Here the MSB 5 bits 10110 represents the integer part We know,

$10110_2 = 1 \times 2^4 + 0 \times 2^3 + 1 \times 2^2 + 1 \times 2^1 + 0 \times 2^0$

$\qquad = 16 + 0 + 4 + 2 + 0$

$\qquad = 22_{10}$

The LSB 3 bits 111 represent fractional part as

$0.1\,11_2 = 1 \times 2^{-1} + 1 \times 2^{-2} + 1 \times 2^{-3}$

$\qquad = \dfrac{1}{2} + \dfrac{1}{4} + \dfrac{1}{8} = \dfrac{4+2+1}{8} = \dfrac{7}{8}$

$\qquad = 0.875_{10}$

∴   10110111 in fixed (8, 3) representation is

$10110.111_2 = 22.875_{10}$ **Ans.**

Q.15. Find the largest and smallest number that can be represented in fixed (8, 3) representation.

Ans. The smallest and largest number comes from the conversion of all 8 bits being 0 and all 8 bits being 1.

∴   The lowest value in fixed (8, 3) is

$00000.000_2 = 0_{10}$

The highest value in fixed (8, 3) representation is

$11111.111_2 = 1 \times 2^4 + 1 \times 2^3 + 1 \times 2^2 + 1 \times 2^1 + 1 \times 2^0 + \dfrac{1}{2} + \dfrac{1}{4} + \dfrac{1}{8}$

$\qquad = 16 + 8 + 4 + 2 + 1 + \dfrac{4+2+1}{8}$

$\qquad = 31 + \dfrac{7}{8} = 31.875_{10}$ **Ans.**

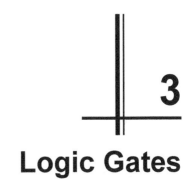

3

Logic Gates

FACTS THAT MATTER

Gate:	A digital circuit with one or more input voltages but only one output voltage.
Basic gates:	There are three basic gates—NOT gate, OR gate and AND gate.
NOT gate:	NOT gate or inverter is a gate with only one input ad a complemented output.
OR gate:	A gate with two or more inputs whose output is high when any or all the inputs are high.
AND gate:	A gate with two or more inputs whose output is high only when all the inputs are high.
Universal gate:	There are two universal gates NAND and NOR. NAND and NOR gates are called universal gates because all the three basic gates can be realised from these gates.
Boolean algebra:	It is the foundation of logic circuits.
Boolean expression:	For NOT gate $\rightarrow Y = \overline{A}$. where Y is the output and A is input. for OR gate $\rightarrow Y = A + B$ where Y is the output and A and B are inputs for AND gate $\rightarrow Y = AB$ where Y is the output and A and B are inputs

for NAND gate → $Y = \overline{AB}$ where Y is the output and A and B are inputs

for NOR gate → $Y = \overline{A+B}$ where Y is the output and A and B are inputs

NAND gate: Means not AND. NAND gate is a gate whose output is high when any or all the inputs are low.

NOR gate: Means not OR. NOR gate is a gate whose output is high only when all the inputs are low.

Exclusive OR gate
or X-OR gate: A gate whose output is high when odd number of inputs are high.

Exclusive NOR gate: A gate that produces a high output for both inputs high or both inputs low.

Symbols and Boolean Expression of gates

Two input AND gate

A ─┐
   ├──── Y $Y = AB$
B ─┘   Boolean expression
   Symbol

Two input OR gate

A ─┐
   ├── Y $Y = A + B$
B ─┘   Boolean expression
   Symbol

NOT gate

A ──▷o── Y $Y = \overline{A}$
                  Boolean expression
   Symbol

Two input NAND gate

A ─┐
   ├o── Y $Y = \overline{AB}$
B ─┘   Boolean expression
   Symbol

Two input NOR gate

A ─┐
   ├o── Y $Y = \overline{A + B}$
B ─┘   Boolean expression
   Symbol

Exclusive OR gate

A ─┐
   ├o── Y $Y = A \oplus B = \overline{A}B + A\overline{B}$
B ─┘   Boolean expression
   Symbol

Truth table: A table that shows all the input-output possibilities of a logic circuit.

Logic circuit: A digital circuit, a switching circuit or any kind of two state circuit that duplicates mental processes.

Positive logic: A system in which high = 1 and low = 0 is known as positive logic.

Negative logic: A system in which high = 0 and low = 1 is known as negative logic.

Bubbled AND gate: AND gate with inverted inputs

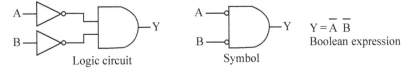

$$Y = \overline{A}\ \overline{B}$$
Boolean expression

Logic circuit   Symbol

Bubbled OR gate: OR gate with inverted inputs

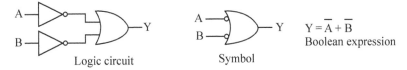

$$Y = \overline{A} + \overline{B}$$
Boolean expression

Logic circuit   Symbol

Assert: To activate. If an input line has a bubble on it, you assert the input by making it low. If there is no bubble, you assert the input by making it high.

AND-OR-Invert gate (AOI):   An integrated circuit containing combinational logic consisting of several AND gates feeding int an OR gate and then an inverter

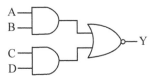

AND-OR-Invert network

The AOI network is available as a separate TTL gate. For *e.g.*, the 7451 is a dual 2-input 2-wide AOI gate meaning two networks as shown above in a single 14-pin TTL package. Here 2-wide means two AND gates across.

Expandable

AND-OR-Invert gates: The widest AOI gate available in the 7400 series is 4-wide. If we need 6 or 8-wide circuit we use an expandable AOI gate. There are two additional inputs bubble and arrow in such type of AOI gate. The output of the expander is connected to bubble and arrow

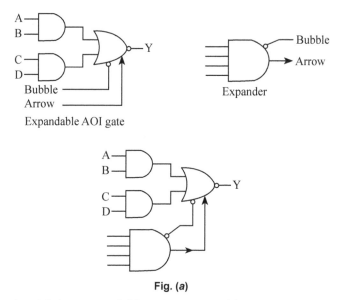

Fig. (a)

Expander driving expandable AOI gate—This means that the expander outputs are being ORed with the signals of the AOI gate. In other words Fig. (a) is equivalent to the AOI circuit of Fig. (b) as shown below

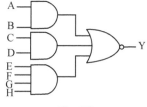

Fig. (b)

De-Morgan's First Theorem: It states that complement of a logical sum is equal to the logical products f the complements. The Boolean equation is $\overline{A + B} = \overline{A} \cdot \overline{B}$.

De-Morgan's Second Theorem: It states that complement of a logical product is equal to the logical sum of the complements.

The Boolean equation is

$$\overline{AB} = \overline{A} + \overline{B}.$$

Duty Cycle is a convenient measure of how symmetrical or how unsymmetrical a waveform is. For a periodic digital signal, the ratio of high level time to the period or the ratio of low level time to the period is known as **Duty Cycle**.

Duty cycle $H = \dfrac{t_H}{T}$

Duty cycle $L = \dfrac{t_L}{T}$

∴ Duty cycle for a symmetrical wave is

$$\text{Duty cycle H = Duty cycle } L = \frac{T/2}{T} = 0.5 \text{ or } 50\%$$

OBJECTIVE TYPE QUESTIONS

1. The expression AB represents _____ gate.
2. NOT gate is also called _____.
3. Exclusive OR gate has high output when _____ number of inputs are high.
4. An OR gate has 6 inputs, then there will be _____ input/output conditions in its truth table.
5. NOR gate is equivalent to OR gate followed by _____ gate.
6. AND, OR and NOT gates are known as _____ gates.
7. NAND and NOR gates are known as _____ gates.
8. How many NAND gates are required to form a two input OR gate?
9. For positive logic higher voltage level is taken as logic _____.
10. A certain digital circuit is designed to operate with voltage levels of –0.2V dc and –3.0v dc. If H = 1 = –0.2 V dc and L = 0 = –3.0 V dc, is it positive logic or negative logic?
11. Write the Boolean expression for an OR gate having A and B as inputs and Y as output.
12. Write an expression for an inverter or NOT gate equivalent to Y = not A.
13. Write the Boolean expression for a 2 input NOR gate.
14. When we speak of an AND-OR-Invert gate, what is the meaning fo 2-wide?
15. What is the purpose of using an expander with an AND-OR-Invert gate?
16. In negative logic binary 0 stands for high voltage and binary 1 stands for _____.
17. Assertion level logic means _____.

18. If a signal causes something to happen when low, it is drawn with a _____, this is an active low signal.

19. In an active high signal, a signal causes something to happen when _____, it is drawn _____.

20. NAND and NOR gates are called universal gates because _____.

Answers

1. AND.

2. Inverter.

3. Odd.

4. 2^6.

5. NOT.

6. Basic.

7. Universal.

8. Three.

9. High.

10. Positive logic.

11. $Y = A + B$.

12. $Y = \overline{A}$.

13. $Y = \overline{A+B}$, **14.** There are two AND gates at the input.

15. It is used to increase the number of input AND gates.

16. Low voltage.

17. Drawing logic symbols to indicate the action of each signal. For active low signal it is drawn with a bubble and for active high, it is drawn without a bubble.

18. Bubble.

19. High, without a bubble.

20. Both can be used individually to represent all the three basic gates— AND, OR and NOT.

SHORT ANSWER TYPE QUESTIONS

Q.1. Derive the truth table of NOT gate.

Ans. In Boolean algebra variable can be either 0 or 1. The output Y of NOT gate is always complement of input A *i.e.* $Y = \overline{A}$

$\therefore$ If $A = 0$, $Y = \overline{0} = 1$

$A = 1$, $Y = \overline{1} = 0$

∴   Truth table is

Input A	Output Y
0	1
1	0

Q.2. Derive the truth table of 2 input OR gate

Ans.

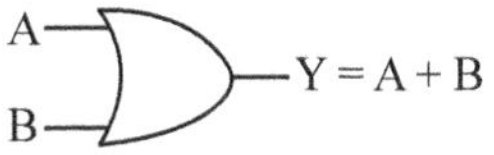

$$A, B \longrightarrow Y = A + B$$

According to definition, the output of 2 input OR gate is high if either or both inputs are high.

Let us consider all the possible cases

If A = 0, B = 0, Y = 0 + 0 = 0

A = 0, B = 1, Y = 0 + 1 = 1

A = 1, B = 0, Y = 1 + 0 = 1

A = 1, B = 1, Y = 1 + 1 = 1

(The '+' sign here represents logic operation OR and not addition operation of basic arithmetic)

∴   The truth table of OR gate

Input		Output
A	B	Y
0	0	0
0	1	1
1	0	1
1	1	1

Q.3. Derive the truth table of 3 input OR gate

Ans.

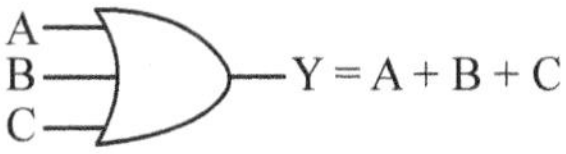

$$A, B, C \longrightarrow Y = A + B + C$$

Logic symbol of 3 I/P OR gate

According to definition, the output of 3 input OR gate is high if any or all the inputs are high.

Let us consider all the possible cases

If A = 0, B = 0, C = 0, Y = 0 + 0 + 0 = 0

A = 0, B = 0, C = 1, Y = 0 + 0 + 1 = 1

A = 0, B = 1, C = 0, Y = 0 + 1 + 0 = 1

$A = 0, B = 1, C = 1, Y = 0 + 1 + 1 = 1$

$A = 1, B = 0, C = 0, Y = 1 + 0 + 0 = 1$

$A = 1, B = 0, C = 1, Y = 1 + 0 + 1 = 1$

$A = 1, B = 1, C = 0, Y = 1 + 1 + 0 = 1$

$A = 1, B = 1, C = 1, Y = 1 + 1 + 1 = 1$

(The '+' sign here represents logic operation OR and not addition operation of basic arithmetic)

∴   The truth table of 3 input OR gate is

Input			Output
A	B	C	Y
0	0	0	0
0	0	1	1
0	1	0	1
0	1	1	1
1	0	0	1
1	0	1	1
1	1	0	1
1	1	1	1

Q.4. Derive the truth table of 2 input AND gate

Ans.

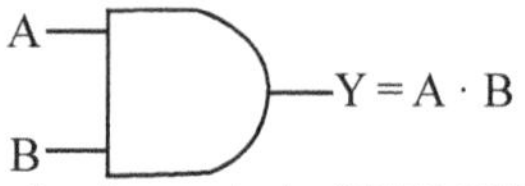

Logic symbol of 2 I/P AND gate

According to definition, the output of 2 I/P AND gate is high only when both the inputs are high.

Let us consider all the possible cases

If $A = 0, B = 0, Y = 0.0 = 0$

$A = 0, B = 1, Y = 0.1 = 0$

$A = 1, B = 0, Y = 1.0 = 0$

$A = 1, B = 1, Y = 1.1 = 1$

∴   The Truth Table of 2 input AND gate is

Input		Output
A	B	Y
0	0	0
0	1	0
1	0	0
1	1	1

Q.5. Derive the truth table of 3 input AND gate

Ans.

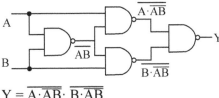

$$Y = A \cdot B \cdot C$$

Logic symbol of 3 I/P AND gate

According to definition, the output of 3 input AND gate is high only when all the inputs are high.

Let us consider all the possible cases

If A = 0, B = 0, C = 0, Y = 0.0.0 = 0

A = 0, B = 0, C = 1, Y = 0.0.1 = 0

A = 0, B = 1, C = 0, Y = 0.1.0 = 0

A = 0, B = 1, C = 1, Y = 0.1.1 = 0

A = 1, B = 0, C = 0, Y = 1.0.0 = 0

A = 1, B = 0, C = 1, Y = 1.0.1 = 0

A = 1, B = 1, C = 0, Y = 1.1.0 = 0

A = 1, B = 1, C = 1, Y = 1.1.1 = 1

∴   The Truth Table of 3 input AND gate is

Input			Output
A	B	C	Y
0	0	0	0
0	0	1	0
0	1	0	0
0	1	1	0
1	0	0	0
1	0	1	0
1	1	0	0
1	1	1	1

Q.6. Implement a two input X-OR gate using exclusively by NAND gates

Ans.

$$Y = \overline{A \cdot \overline{AB}} \cdot \overline{B \cdot \overline{AB}}$$

$$= \overline{\overline{A \cdot \overline{AB}}} + \overline{\overline{B \cdot \overline{AB}}} \text{ (using DeMorgan's 2nd law)}$$

$$= A \cdot \overline{AB} + B \cdot \overline{AB} \ (\because \overline{\overline{A}} = A)$$

$= A \cdot (\overline{A} + \overline{B}) + B \cdot (\overline{A} + \overline{B})$ (using DeMorgan's 2nd law)

$= A \cdot \overline{A} + A\overline{B} + B \cdot \overline{A} + B\overline{B}$

$= 0 + A\overline{B} + \overline{A}B + 0$ $(\because A\overline{A} = 0)$

$= \overline{A}B + A\overline{B}$ which is the Boolean expression for X-OR gate

Thus two input X-OR gate is implemented using only NAND gates.

Q.7. Implement a two input X-OR gate using exclusively by NOR gates

Ans.

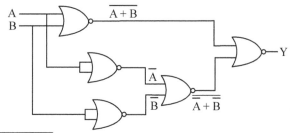

$Y = \overline{\overline{A+B} + \overline{\overline{A}+\overline{B}}}$

$= \overline{\overline{A+B}} \cdot \overline{\overline{\overline{A}+\overline{B}}}$ (using DeMorgan's 1st law)

$= (A+B)(\overline{A}+\overline{B})$ $(\because \overline{\overline{A}} = A)$

$= A\overline{A} + A\overline{B} + B\overline{A} + B\overline{B}$

$= 0 + A\overline{B} + \overline{A}B + 0$ $(\because A\overline{A} = 0)$

$= \overline{A}B + A\overline{B}$ which is the Boolean expression for X-OR gate

Thus two input X-OR gate is implemented using only NOR gates.

Q.8. Define ex-OR gate and derive its truth table.

Ans. Ex-OR gate is a gate whose output is high when odd number of inputs are high.

Basic logic circuit for Ex-OR gate is shown below:

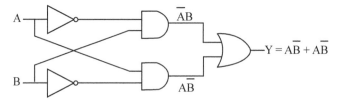

Let us consider all the input conditions

If $A = 0$, $B = 0$, $Y = \overline{0}.0 + 0.\overline{0} = 1.0 + 0.1 = 0 + 0 = 0$

If $A = 0$, $B = 1$, $Y = \overline{0}.1 + 0.\overline{1} = 1.1 + 0.0 = 1 + 0 = 1$

If $A = 1$, $B = 0$, $Y = \overline{1}.0 + 1.\overline{0} = 0.0 + 1.1 = 0 + 1 = 1$

If $A = 1$, $B = 1$, $Y = \overline{1}.1 + 1.\overline{1} = 0.1 + 1.0 = 0 + 0 = 0$

The truth table of two input ex-OR gate is as follows:

Input		Output
A	B	Y
0	0	0
0	1	1
1	0	1
1	1	0

Q.9. What is X-NOR gate. Draw the symbol and derive its truth table.

Ans. Ex-NOR gate is a gate that produces a high output for both inputs high or both inputs low.

$$Y = \overline{\overline{AB} + A\overline{B}}$$

A —⟩D⟩o— Y = $\overline{AB} + A\overline{B}$
B —

Basic circuit for X-NOR gate is shown below:

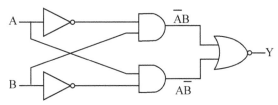

$$Y = \overline{\overline{A}B + A\overline{B}} = \overline{\overline{A}B} \cdot \overline{A\overline{B}} \quad \text{(using De-Morgan's 1st law)}$$
$$= (A + \overline{B})(\overline{A}+B) \quad \text{(using De-Morgan's 2nd law)}$$
$$= A\overline{A} + AB + \overline{A}\,\overline{B} + \overline{B}B$$
$$= 0 + AB + \overline{A}\,\overline{B} + 0 \ (\because A\overline{A} = 0)$$
$$= AB + \overline{A}\,\overline{B}$$

$\therefore$ $Y = AB + \overline{A}\,\overline{B}$ is the Boolean expression for X-NOR gate.

Now let us consider all the possible inputs.

If A = 0, B = 0, Y = 0.0 + $\overline{0}.\overline{0}$ = 0 + 1.1 = 0 + 1 = 1

If A = 0, B = 1, Y = 0.1 + $\overline{0}.\overline{1}$ = 0 + 1.0 = 0 + 0 = 0

If A = 1, B = 0, Y = 1.0 + $\overline{1}.\overline{0}$ = 0 + 0.1 = 0 + 0 = 0

If A = 1, B = 1, Y = 1.1 + $\overline{1}.\overline{1}$ = 1 + 0.0 = 1 + 0 = 1

$\therefore$ Thus truth table is as shown below

Input		Output
A	B	Y
0	0	1
0	1	0
1	0	0
1	1	1

Q.10. What is the output Y for the logic circuit shown below:

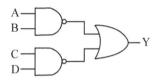

Ans.

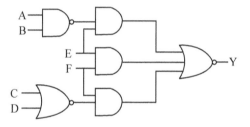

$\therefore$ Output $Y = \overline{AB} + \overline{CD}$

Q.11. What is the output Y for the given logic circuit?

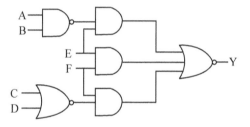

Ans.

$E(\overline{AB}) = E(\overline{A+B})$ (using De-Morgan's 2nd law)

$F\overline{(C+D)} = F\overline{C}.\overline{D}$ (using De-Morgan's 1st law)

$\therefore$ $Y = \overline{E\overline{A} + E\overline{B} + EF + F\overline{C}\,\overline{D}}$

$= \overline{E\,(\overline{A} + \overline{B} + F) + F\overline{C}\,\overline{D}}$

$= \overline{E\,(A + B + F)} \cdot \overline{(F\overline{C}\,\overline{D})}$ (using De-Morgan's 1st law)

$= \overline{E} + \overline{(A + B + F)} \cdot (\overline{F} + C + D)$ (using De-Morgan's 2nd law)

$= (\overline{E} + AB\overline{F})\,(\overline{F} + C + D)$ (using De-Morgan's 1st law)

$\therefore$ $Y = (\overline{E} + AB\overline{F})\,(\overline{F} + C + D)$ **Ans.**

Q.12. Draw the logic circuit for Boolean expression

$$Y = \overline{ABC} = A\overline{B}C + AB\overline{C} + \overline{A}B\overline{C}$$

Ans.

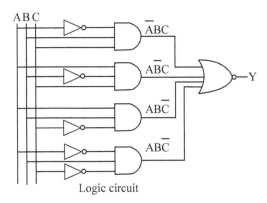

Logic circuit

$$Y = \overline{A}BC + A\overline{B}C + AB\overline{C} + \overline{A}B\overline{C}$$

LONG ANSWER TYPE QUESTIONS

Q.1. Draw the logic circuit for Boolean expression

$$Y = \overline{A + B} + \overline{C}$$ and derive its truth table

Ans. Logic circuit for $Y = \overline{A + B} + \overline{C}$ is shown below:

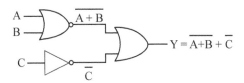

To derive the truth table we have to find out all the input/output possibilities. Since there are three inputs ∴ there will be $2^3 = 8$ input conditions.

(*i*) If A = 0, B = 0, C = 0, then $Y = \overline{0 + 0} + \overline{0} = \overline{0} + 1 = 1 + 1 = 1$

(*ii*) A = 0, B = 0, C = 1, then $Y = \overline{0 + 0} + \overline{1} = \overline{0} + 0 = 1 + 0 = 1$

(*iii*) A = 0, B = 1, C = 0, then $Y = \overline{0 + 1} + \overline{0} = \overline{1} + 1 = 0 + 1 = 1$

(*iv*) A = 0, B = 1, C = 1, then $Y = \overline{0 + 1} + \overline{1} = \overline{1} + 0 = 0 + 0 = 0$

(*v*) A = 1, B = 0, C = 0, then $Y = \overline{1 + 0} + \overline{0} = \overline{1} + 1 = 0 + 1 = 1$

(*vi*) A = 1, B = 0, C = 1, then $Y = \overline{1 + 0} + \overline{1} = \overline{1} + 0 = 0 + 0 = 0$

(*vii*) A = 1, B = 1, C = 0, then $Y = \overline{1 + 1} + \overline{0} = \overline{1} + 1 = 0 + 1 = 1$

(*viii*) A = 1, B = 1, C = 1, then $Y = \overline{1 + 1} + \overline{1} = \overline{1} + 0 = 0 + 0 = 0$

∴   The required us truth table is as shown below

Input			Output
A	B	C	Y
0	0	0	1
0	0	1	1
0	1	0	1
0	1	1	0
1	0	0	1
1	0	1	0
1	1	0	1
1	1	1	0

Q.2. State and prove De-Morgan's theorems.

Ans. De-Morgan's First Theorem: It states that complement of a logical sum is equal to the logical products of the complements. The Boolean equation is

$$\overline{A + B} = \overline{A} \cdot \overline{B}$$

Proof: Let us consider all the four possible cases

Case 1: If $A = 0$, $B = 0$

$\text{L.H.S.} = \overline{A + B} = \overline{0 + 0} = \overline{0} = 1$

$\text{R.H.S.} = \overline{A} \cdot \overline{B} = \overline{0} \cdot \overline{0} = 1.1 = 1$

∴   L.H.S. = R.H.S.

Case 2: If $A = 0$, $B = 1$

$\text{L.H.S.} = \overline{A + B} = \overline{0 + 1} = \overline{1} = 0$

$\text{R.H.S.} = \overline{A} \cdot \overline{B} = \overline{0} \cdot \overline{1} = 1.0 = 0$

∴   L.H.S. = R.H.S.

Case 3: If $A = 1$, $B = 0$

$\text{L.H.S.} = \overline{A + B} = \overline{1 + 0} = \overline{1} = 0$

$\text{R.H.S.} = \overline{A} \cdot \overline{B} = \overline{1} \cdot \overline{0} = 0.1 = 0$

∴   L.H.S. = R.H.S.

Case 4: If $A = 1$, $B = 1$

$\text{L.H.S.} = \overline{A + B} = \overline{1 + 1} = \overline{1} = 0$

$\text{R.H.S.} = \overline{A} \cdot \overline{B} = \overline{1} \cdot \overline{1} = 0.0 = 0$

∴   L.H.S. = R.H.S.

Since we find that for all the four possible cases L.H.S. = R.H.S., hence De-Morgan's First Theorem is proved.

De-Morgan's Second Theorem: It states that complement of a logical product is equal to the logical sum of the complements. The Boolean equation is

$$\overline{AB} = \overline{A} + \overline{B}$$

Proof: Let us consider all the four possible cases

Case 1: If A = 0, B = 0

$\qquad$ L.H.S. = $\overline{0 \cdot 0}$ = $\overline{0}$ = 1

$\qquad$ R.H.S. = $\overline{0} + \overline{0}$ = 1 + 1 = 1

$\qquad \therefore$ L.H.S. = R.H.S.

Case 2: If A = 0, B = 1

$\qquad$ L.H.S. = $\overline{0 \cdot 1}$ = $\overline{0}$ = 1

$\qquad$ R.H.S. = $\overline{0} + \overline{1}$ = 1 + 0 = 1

$\qquad \therefore$ L.H.S. = R.H.S.

Case 3: If A = 1, B = 0

$\qquad$ L.H.S. = $\overline{1 \cdot 0}$ = $\overline{0}$ = 1

$\qquad$ R.H.S. = $\overline{1} + \overline{0}$ = 0 + 1 = 1

$\qquad \therefore$ L.H.S. = R.H.S.

Case 4: If A = 1, B = 1

$\qquad$ L.H.S. = $\overline{1 \cdot 1}$ = $\overline{1}$ = 0

$\qquad$ R.H.S. = $\overline{1} + \overline{1}$ = 0 + 0 = 0

$\qquad \therefore$ L.H.S. = R.H.S.

$\qquad$ Since we find for all the four possible cases L.H.S. = R.H.S., hence De-Morgan's Second Theorem is proved.

Q.3. Why NAND and NOR gates are called universal gates?

Ans. NAND and NOR gates are called universal gates because all the three basic gates can be realised using these gates.

$\qquad$ Now, let us see how NAND gate can be used as all the three basic gates

$\qquad$ (*i*) NAND as NOT gate

$\qquad$ A —$\boxed{\text{D}}$— Y = $\overline{A}$

$\qquad$ which is the output of NOT gate

$\qquad$ (*ii*) NAND as AND gate

$\qquad$ A —$\boxed{\text{D}}$— $\overline{AB}$ —$\boxed{\text{D}}$— Y = $\overline{\overline{AB}}$ = AB
$\qquad$ B

$\qquad$ which is the Boolean expression of AND gate

$\qquad$ (*iii*) NAND as OR gate

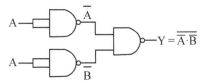

using De-Morgan's 2nd law $Y = \overline{\overline{A} \cdot \overline{B}} = \overline{\overline{A}} + \overline{\overline{B}} = A + B$ which is the Boolean expression for OR gate.

Hence we find that NAND gate can be used as all the three basic gates *i.e.* NOT, OR and AND. Therefore NAND is an universal gate.

Now let us see how NOR gate can be used as all the three basic gates.

(*i*) NOR as NOT gate

A $\longrightarrow$ $Y = \overline{A}$

which is the Boolean expression of NOT gate

(*ii*) NOR as OR gate

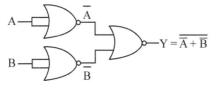

A — B — $\overline{A+B}$ — $Y = \overline{\overline{A+B}} = A + B$

which is the Boolean expression of OR gate

(*iii*) NOR as AND gate

A $\longrightarrow$ $\overline{A}$

B $\longrightarrow$ $\overline{B}$

$Y = \overline{\overline{A} + \overline{B}}$

Using De-Morgan's 1st law $Y = \overline{\overline{A} + \overline{B}} = \overline{\overline{A}} . \overline{\overline{B}} = A . B$ which is the Boolean expression for AND gate.

Hence we find that NOR gate can be used as all the three basic gates *i.e.* NOT, OR and AND. Therefore NOR is an universal gate.

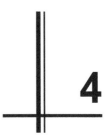

Combinational Logic Circuits and Simplification

4

FACTS THAT MATTER

1. Boolean Laws and Theorems

Basic laws:

The commutative laws are

$$A + B = B + A \tag{1}$$
$$AB = BA \tag{2}$$

The Associative laws are

$$A+(B+C) = (A+B)+C \tag{3}$$
$$A(BC) = (AB) C \tag{4}$$

The Distributive laws are

$$A(B+C) + AB+AC \tag{5}$$
$$A + BC + (A+B) (A + C) \tag{6}$$

Other law and Theorems

$$A + 0 = A \tag{7}$$
$$A.0 = 0 \tag{8}$$
$$A + 1 = 1 \tag{9}$$
$$A \ 1=A \tag{10}$$
$$A + A = A \tag{11}$$
$$A \ A = A \tag{12}$$
$$A + \overline{A} = 1 \tag{13}$$

$$A \cdot \overline{A} = 0 \tag{14}$$
$$\overline{\overline{A}} = A \tag{15}$$
$$A + AB = A \tag{16}$$
$$A(A + B) = A \tag{17}$$
$$A + \overline{A}B = A + B \tag{18}$$
$$A(\overline{A} + B) = AB \tag{19}$$
$$AB + A\overline{B} = A \tag{20}$$
$$(A + B)(A + \overline{B}) = A \tag{21}$$

De-Morgan's laws

$$\overline{A+B} = \overline{A}\,\overline{B} \tag{22}$$
$$\overline{AB} = \overline{A} + \overline{B} \tag{23}$$

Consensus Theorem

$$AB + \overline{A}C + BC = AB + \overline{A}C \tag{24}$$
$$(A+B)(\overline{A} + C)(B+C) = (A+B)(\overline{A}+C) \tag{25}$$

2. **Consensus Theorem:** A theorem that simplifies a Boolean equation removing a redundant consensus term.

3. **Dual circuit:** Given a logic circuit, its dual can be found changing each AND (NAND) gate to an OR (NOR) gate, changing each OR (NOR) gate to an AND (NAND) gate and complement all input-output signals.

4. **Karnaugh map:** A drawing that shows all the fundamental products and the corresponding output values of a truth table.

5. **Pair:** Two horizontally or vertically adjacent 1s on a Karnaugh map.

6. **Quad:** Four horizontal, vertical or rectangular adjacent 1s on a Karnaugh map.

7. **Octet:** Eight adjacent 1s in a 2×4 shape on a Karnaugh map.

8. **Overlapping Groups:** Groups formed using the same 1 more than once when looping the 1s of a Karnaugh map.

9. **Redundant group:** A group of 1 s on a Karnaugh map that are all part of other groups. This redundant group should be eliminated while simplification.

10. **Sum of Products (SOP) Equation:** The logical sum of those fundamental products that produce output 1 s in the truth table. The corresponding logic circuit is an AND-OR circuit or the equivalent NAND-NAND circuit.

11. **Product of sums (POS) Equation:** The logical product of those fundamental sums that produce output 1 s in the truth table. The corresponding logic circuit is an OR-AND circuit or equivalent NOR-NOR circuit.

12. **Literal:** A literal is a variable or its complement

13. **Minterm:** A minterm of n variables is a product of n literals in which each variable appears exactly once in either true or complemented form, but not both.

14. **Standard SOP:** When a function f is written as a sum of minterms, it is referred to as a minterm expansion or a standard sum of products (SOP),

    e.g. $f = A'\,BC + AB'C' + AB'C + ABC' + ABC$

    where each term is a minterm.

    In general minterm is designated by m_i

    $\therefore$ The above equation can be rewritten as

    $f(A, B, C) = m_3 + m_4 + m_5 + m_6 + m_7$

    $\qquad = \Sigma\, m\,(3, 4, 5, 6, 7)$

15. **Maxterm:** A maxterm of n variables is a sum of n literals in which each variable appears exactly once in either true or complemental form, but not both.

16. **Standard POS:** When a function f is written as a product of maxterms, it is refered to as a maxterm expansion or standard product of sums (POS)

    e.g. $f = (A + B + C)\,(A + B + C')\,(A + B' + C)$

    where each term is a maxterm

    In general maxterm is designated by Mi. Further each maxterm has a value of 0 for exactly one combination of values for A, B and C. Thus

    if $A = B = C = 0$, $A + B + C = 0$, if $A = B = 0$, $C = 1$

    $A + B + C' = 0$ etc.

    Notice that each maxterm is complement of the corresponding minterm i.e. $M_i = m'_i$. With M notation, the above equation can be rewritten as

    $f(A,B,C) = M_0\, M_1\, M_2$

    $\qquad = \pi\, M\,(0, 1, 2)$

17. **Don't care condition:** An input - output condition that never occurs during normal operation. Since the condition never occurs, we can use X on Karnaugh map. This X can be a 0 or 1, whichever we prefer.

18. **Quine-McClusky method:** This is a tabular method for logic simplification.

19. **Implicant:** Any single 1 or any group of 1's which can be combined together on a map of the function F represents a product term which is called an implicant of F.

20. **Prime Implicant:** A product term implicant is called a prime implicant if it cannot be combined with another term to eliminate a variable.

21. **Essential Prime Implicant:** If a minterm is covered by only one prime implicant, that prime implicant is said to be essential Prime implicant and it must be included in the minimum sum-of-products.

22. **Hazard:** Unwanted glitches due to finite propagation delay of logic circuit.

23. **Hazard cover:** Additional gates in logic circuit preventing hazard.

OBJECTIVE TYPE QUESTIONS

1. All the rules for Booloan algera are exactly the same as for ordinary algebra (True or False)

2. Expand using the distributive law : $Y = A(B+C)$

3. Simplify : $Y = \overline{A}Q + AQ$

4. How many fundamental products are there for two variables? How many for three variables?

5. Why distributive law of switching algebra comes under basic postulates while commutative law does not?

6. Are Ex-OR and Ex-NOR functions dual of each other?

7. The AND-OR or the NAND-NAND circuit obtained with the sum-of-products method is always the simplest possible circuit (True or False)

8. What is a Karnaugh map?

9. How many entries are there on a four-variable Karnaugh map?

10. On a Karnaugh map, two adjacent 1 s are called a _____.

11. On a Karnaugh map, an octet contains how many $1s$?

12. A quad on a Karnaugh map contains _____ $1s$.

13. What is meant by a don't - care condition on a Karnaugh map? How is it indicated?

14. How can using don't - cares aid circuit simplification?

15. A product-of-sums expression leads to what kind of logic circuit?

16. Explain how to convert the complementary NAND-NAND circuit into its dual NOR-NOR circuit.

17. What is a prime implicant?

18. What are the advantages of tabular method?

19. What is static - o hazard?

20. What is dynamic hazard?

21. The expression $AB + A\overline{B} + A\overline{C} + \overline{A}\ \overline{C} = ?$

22. The expression $\overline{A}B + A\overline{B} + AB + ?$

Answers

1. False.

2. $Y = AB + AC$.

3. $Y = Q$.

4. Four, Eight.

5. Because the commutative law can be derived from others.

6. Yes.

7. False.

8. A Karnaugh map is a visual display of all the fundamental products and the corresponding output values of a truth table.

9. Sixteen.

10. Pair.

11. Eight.

12. Four.

13. A don't care condition is an input condition that never occurs during normal operations. It is indicated with an X on a Karnaugh map.

14. An X can be used to create pairs, quads, octets etc.

15. A product-of-sums expression leads directly to an OR-AND circuit.

16. Change all NAND gates to NOR gates and complement all signals.

17. Prime implicant is a product term implicant if it cannot be combined with another term to eliminate a variable.

18. Systematic, step-by-step approach that can be implemented in a digital computer and providing solution for any number of variables.

19. A logic high pulse of very short duration when output should be at logic low.

20. Dynamic hazard occurs when circuit output makes multiple transitions before it settles while the logic equation asks for only one transition.

21. $A + \overline{C}$.

22. A+B.

SHORT ANSWER TYPE QUESTIONS

Q.1. Simplify the Boolean equation $Y = \overline{A}\,\overline{B}\,\overline{C} + \overline{A}\,B\,\overline{C} + A\,\overline{B}\,\overline{C} + AB\,\overline{C}$ and draw the simplified logic circuit

Ans. $Y = \overline{A}\,\overline{B}\,\overline{C} + \overline{A}\,B\,\overline{C} + A\,\overline{B}\,\overline{C} + AB\,\overline{C}$

$\quad = \overline{C}\,(\overline{A}\,\overline{B} + \overline{A}B + A\overline{B} + AB)$ using distributive 1st law

$\quad = \overline{C}\,[\overline{A}\,(\overline{B} + B) + A\,(\overline{B} + B)]$ using distributive 1st law

$\quad = \overline{C}\,[\overline{A}(1) + A\,(1)]$ $\qquad \because A + \overline{A} = 1$

$\quad = \overline{C}\,(\overline{A} + A) = \overline{C}\,1$ $\qquad \because A + \overline{A} = 1$

$\quad \therefore\ Y = \overline{C}$

The simplified equation means we don't even need a logic circuit. All we need is a wire connecting input $\overline{C}$ to output Y.

Q.2. Prove that $A(A' + C)\,(A'B + C)\,(A'BC + C') = 0$

Ans. L.H.S. $A(A' + C)\,(A'B + C)\,(A'BC + C')$

$\quad = (AA' + AC)\,(A'B + C)\,(A'BC + C')$ using distributive law

$\quad = AC\,(A'B + C)\,(A'BC + C')$ $\qquad \because AA' = 0$

$\quad = (AC.A'B + AC.C)\,(A'BC + C')$ using distributive law

$\quad = (0 + AC)\,(A'BC + C')$ $\qquad \because AA' = 0, A \cdot A = A$

$\quad = AC\,(A'BC + C')$

$\quad = AC.A'BC + AC.C'$ $\qquad$ using distributive law

$\quad = 0 + 0$ $\qquad \because AA' = 0$

$\quad = 0 = $ R.H.S.

Hence proved.

Q.3. Simplify the Boolean expression

$Y = (A + B)\,\overline{(A'(B' + C'))} + A'(B+C)$ using Boolean laws and Theorems and draw the simplified logic circuit.

Ans. $Y(A + B)\,\overline{(A'(B'+C'))} + A'\,(B + C)$

$\quad = (A + B)\,(\overline{A} + \overline{(B'+C')}) + A'(B + C)$ using De Morgan's 2nd law

$\quad = (A + B)\,(A + (\overline{B}\,\overline{C})) + A'(B + C)$ using De Morgan's 1st law

$\quad = (A + B)\,(A + BC) + A'(B + C)$

$$= (AA + AB + ABC + BBC) + A'(B + C)$$
$$= (A + AB + ABC + BC) + A'(B + C) \qquad \because A\,A = A$$
$$= A(1 + B + BC) + BC + A'\,(B + C)$$
$$= A + BC + A'(B + C) \qquad\qquad\quad \because 1 + B + BC = 1$$
$$= A + A'\,(B + C) + BC$$
$$= A + B + C + BC \qquad\qquad\qquad \because A + A'B = A + B$$
$$= A + B + C\,(1 + B)$$
$$= A + B + C \qquad\qquad\qquad\qquad\quad \because 1 + B + 1$$

The simplified logic circuit is shown below:

A —⟍
B — ⟩— Y=A+B+C
C —⟋

Q.4. Simplify the Boolean expression $\overline{\overline{X\overline{Y}} + XYZ} + X(Y + X\overline{Y})$ using Boolean laws & Theorems.

Ans. $F = \overline{\overline{X\overline{Y}} + XYZ} + X(Y + X\overline{Y})$

$= (\overline{\overline{X\overline{Y}} + XYZ}) \cdot \overline{X\,(Y + X\overline{Y})}$ using De-Morgan's 1st law

$= (X\overline{Y} + \overline{XYZ})\,(\overline{X} + \overline{Y + X\overline{Y}})$ using De-Morgan's 2nd law

$= (X\overline{Y} + \overline{XYZ})\,\{\overline{X} + \overline{Y}\,(\overline{X\overline{Y}})\}$ using De-Morgan's 1st law

$= (X\overline{Y} + XYZ)\,\{\overline{X} + \overline{Y}\,(\overline{X} + Y)\}$ using De-Morgan's 2nd law

$= (X\overline{Y} + XYZ)\,(\overline{X}(1 + \overline{Y}) + 0)$ $\because Y\,\overline{Y} = 0$

$= (X\overline{Y} + XYZ)\,\overline{X}$ $\because 1 + \overline{Y} = 0$

$= X\,\overline{X}\,\overline{Y} + X\,\overline{X}YZ$

$= 0 + 0$ $\because X\overline{X} = 1$

$= 0.$

Q.5. Simplify the Boolean expression $XY + XYZ + \overline{X}Y + X\overline{Y}Z$ using Boolean laws and Theorems.

Ans. $F = XY + XYZ + \overline{X}Y + X\overline{Y}Z$

$= XY + \overline{X}Y + XYZ + X\overline{Y}Z$

$= Y\,(X + \overline{X}) + XZ\,(Y + \overline{Y})$

$= Y.1 + XZ1$ $\because X + \overline{X} = 1$

$= Y + XZ.$

Q.6. Using Boolean Algebra, simplify the following circuit to get a single gate output.

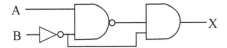

Ans.

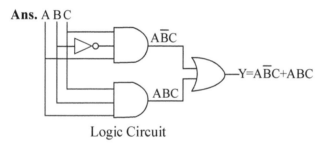

$X = \overline{\overline{AB} \ \overline{B}}$

$= (\overline{\overline{A}} + \overline{\overline{B}}) \, . \, \overline{B}$ using De-Morgen's 2nd law

$= (\overline{A} + B) \, \overline{B}$

$= \overline{A} \, \overline{B} + B\overline{B}$ using distributive law

$= \overline{A} \, \overline{B} + 0$ $\therefore B\overline{B} = 0$

$= \overline{A} \, \overline{B}$

Hence the single gate output is as follows:

A ————o⟩
B ————o⟩——— $X = \overline{A} \, \overline{B}$

Q.7. Draw the logic circuit for Y=A$\overline{B}$C + ABC. Next, simplify the equation with Boolean algebra and draw the simplified logic circuit.

Ans. A B C

A$\overline{B}$C

ABC

—Y=A$\overline{B}$C+ABC

Logic Circuit

Now let us simplify

Y= A$\overline{B}$C + ABC

= AC ($\overline{B}$ + B) using distributive law

= AC . 1 $\because \overline{B} + B = 1$

= AC

$\therefore$ The simplified circuit will be as follows

A ————⟩
C ————⟩——— Y= AC

Q.8. Minimise the expression Y = F(A, B, C) = $\overline{A} \, \overline{B} \, \overline{C} + \overline{A} \, \overline{B} \, C + \overline{A}BC + A\overline{B}C$ using Boolean laws and Theorems.

Ans. Y= $\overline{A} \, \overline{B} \, \overline{C} + \overline{A} \, \overline{B} \, C + \overline{A} \, B \, C + A \, \overline{B} \, C$

Since in Boolean algebra X=X+X+X+, extending the given equation we can write

$$Y = \overline{A}\,\overline{B}\,\overline{C} + (\overline{A}\,\overline{B}\,C + \overline{A}\,\overline{B}\,C + \overline{A}\,\overline{B}\,C) + \overline{A}\,B\,C + A\,\overline{B}\,C$$
$$= (\overline{A}\,\overline{B}\,\overline{C} + \overline{A}\,\overline{B}\,C) + (\overline{A}\,\overline{B}\,C + \overline{A}\,B\,C) + (\overline{A}\,\overline{B}\,C + A\,\overline{B}\,C)$$

using associative law

$$= \overline{A}\,\overline{B}\,(\overline{C}+C) + \overline{A}C\,(\overline{B} + B) + \overline{B}C\,(\overline{A} + A)$$

using distributive law

$$= \overline{A}\,\overline{B}.1 + \overline{A}C.1 + \overline{B}C.1 \qquad \therefore\ X + X'=1$$
$$= \overline{A}\,\overline{B} + \overline{A}C + \overline{B}C \qquad\quad \therefore\ X.1 = X$$

Q.9. Minimize the expression $Y = F(A,B,C) = \overline{A}\,\overline{B}\,\overline{C} + \overline{A}\,\overline{B}\,C + \overline{A}\,BC + A\overline{B}C$ using Karnaugh map.

Ans. $Y = F(A,B,C) = \overline{A}\,\overline{B}\,\overline{C} + \overline{A}\,\overline{B}\,C + \overline{A}\,B\,C + A\,\overline{B}\,C$

To simplify using Karnaugh map, first we have to write the truth table consisting of inputs A,B,C and output Y=1 for the fundamental products mentioned in the given equation. The truth table is shown below.

A	B	C	Y	Fundamental Product
0	0	C	1	$\overline{A}\,\overline{B}\,\overline{C}$
0	0	1	1	$\overline{A}\,\overline{B}\,C$
0	1	0	0	
0	1	1	1	$\overline{A}\,B\,C$
1	0	0	0	
1	0	1	1	$A\,\overline{B}\,C$
1	1	0	0	
1	1	1	0	

Now let us draw the Karnaugh map and place 1 for the corresponding fundamental products mentioned in the truth table and fill the remaining places with 0.

$$\therefore\ Y = \overline{A}\,\overline{B} + \overline{A}\,C + \overline{B}\,C$$

Q.10. Simplify the function $F(A,B,C) = \Sigma\, m(0,2,4,5,6)$ using Karnaugh map.

Ans. $F(A,B,C) = \Sigma\, m(0,2,4,5,6)$
$$= m_0 + m_2 + m_4 + m_5 + m_6$$
$$= \overline{A}\,\overline{B}\,\overline{C} + \overline{A}\,B\,\overline{C} + A\,\overline{B}\,\overline{C} + A\,\overline{B}\,C + AB\overline{C}$$

To simplify using Karnaugh map, first let us write the truth table consisting of inputs A,B,C and output Y=1 for the fundamental products mentioned in the given equation. The truth table is shown below:

A	B	C	Y	Fundamental Product
0	0	0	1	$\overline{A}\,\overline{B}\,\overline{C}$
0	0	1	0	
0	1	0	1	$\overline{A}\,B\,\overline{C}$
0	1	1	0	
1	0	0	1	$A\,\overline{B}\,\overline{C}$
1	0	1	1	$A\,\overline{B}\,C$
1	1	0	1	$A\,B\,\overline{C}$
1	1	1	0	

Now let us draw the Karnaugh map and place 1 for the corresponding fundamental products mentioned in the truth table and fill the remaining places with 0.

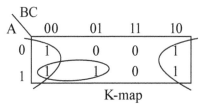

K-map

$$\therefore\ Y = \overline{C} + A\overline{B}\ \textbf{Ans.}$$

Q.11. Minimize the following Boolean expression using K-map

F(A, B, C, D) = $\Sigma\,m$(0, 1, 3, 5, 6, 9, 11, 12, 13, 15)

Ans. F= $\Sigma\,m$ (0, 1, 3, 5, 6, 9, 11, 12, 13, 15)

$= m_0 + m_1 + m_3 + m_5 + m_6 + m_9 + m_{11} + m_{12} + m_{13} + m_{15}$

$= \overline{A}\,\overline{B}\,\overline{C}\,\overline{D} + \overline{A}\,\overline{B}\,\overline{C}D + \overline{A}\,\overline{B}CD + \overline{A}\,B\overline{C}D + \overline{A}\,BC\overline{D}$

$+ A\,\overline{B}\,\overline{C}\,D + A\overline{B}CD + AB\,\overline{C}\,\overline{D} + AB\,\overline{C}\,D + ABCD$

To simplify using Karnaugh map we have to place 1 in the map for the corresponding fundamental products given in equation and fill the remaining places of the map with 0.

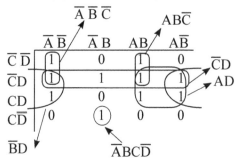

$$\therefore \ Y = AD + \overline{C}D + \overline{B}D + \overline{A}\,\overline{B}\,\overline{C} + AB\,\overline{C} + \overline{A}BCD \quad \textbf{Ans.}$$

Q.12. Making use of Karnaugh mapping procedure simplify the following expression.

$$Y = A\overline{B} + \overline{A}\,\overline{B}C + AB\overline{C}D + \overline{A}\,\overline{B}\,\overline{C} + A\,\overline{C}\,\overline{D}$$

Ans. $Y = A\overline{B} + \overline{A}\,\overline{B}C + AB\overline{C}D + \overline{A}\,\overline{B}\,\overline{C} + A\,\overline{C}\,\overline{D}$

To make use of Karnaugh map, first we have to get the standard S.O.P for the given expression i.e. each term should be a minterm.

$$\therefore \ Y = A\overline{B}\,(C + \overline{C})\,(D + \overline{D}) + \overline{A}\,\overline{B}\,C(D + \overline{D}) + AB\overline{C}D + \overline{A}\,\overline{B}\,\overline{C}\,(D + \overline{D})$$
$$+ A\,\overline{C}\,\overline{D}\,(B + \overline{B}) \quad (\because X + \overline{X} = 1)$$
$$=(A\overline{B}C + A\overline{B}\,\overline{C})(D + \overline{D}) + \overline{A}\,\overline{B}CD + \overline{A}\,\overline{B}C\overline{D} + AB\overline{C}D + \overline{A}\,\overline{B}\,\overline{C}D + \overline{A}\,\overline{B}\,\overline{C}\,\overline{D}$$
$$+ AB\overline{C}\,\overline{D} + A\,\overline{B}\,\overline{C}\,\overline{D}$$
$$= A\overline{B}CD + A\overline{B}C\overline{D} + A\overline{B}\,\overline{C}D + A\overline{B}\,\overline{C}\overline{D} + \overline{A}\,\overline{B}CD + \overline{A}\,\overline{B}C\overline{D}$$
$$+ AB\overline{C}D + \overline{A}\,\overline{B}\,\overline{C}D + \overline{A}\,\overline{B}\,\overline{C}\,\overline{D} + AB\overline{C}\,\overline{D} + A\overline{B}\overline{C}\overline{D} \quad (\because X + X = X)$$

Now let us draw the Karnaugh map by placing 1 for the corresponding minterms or fundamental products given in the equation and fill the remaining places by 0.

	$\overline{A}\,\overline{B}$	$\overline{A}\,B$	AB	$A\overline{B}$
$\overline{C}\,\overline{D}$	1	0	1	1
$\overline{C}D$	1	1	1	1
CD	1	0	0	1
$C\overline{D}$	1	0	0	1

$$\therefore \ Y = \overline{B} + A\,\overline{C} \quad \textbf{Ans.}$$

Q.13. Minimize and implement using NAND logic

$$F = \Sigma m\ (0, 1, 2, 3, 10, 11, 12, 13, 14, 15)$$

Ans. We will minimize using Karnaugh map

$$F = \Sigma m\ (0, 1, 2, 3, 10, 11, 12, 13, 14, 15)$$
$$= m_0 + m_1 + m_2 + m_3 + m_{10} + m_{11} + m_{12} + m_{13} + m_{14} + m_{15}$$
$$= \overline{A}\,\overline{B}\,\overline{C}\,\overline{D} + \overline{A}\,\overline{B}\,C\overline{D} + \overline{A}\,\overline{B}CD + \overline{A}\,\overline{B}C\overline{D} + A\,\overline{B}C\overline{D} + A\,\overline{B}CD$$
$$+ AB\overline{C}\,\overline{D} + AB\overline{C}D + ABC\overline{D} + ABCD$$

Now let us draw the Karnaugh map

	$\overline{C}\,\overline{D}$	$\overline{C}\,D$	CD	$C\overline{D}$
$\overline{A}\,\overline{B}$	1	1	1	1
$\overline{A}B$	0	0	0	0
AB	1	1	1	1
$A\overline{B}$	0	0	1	1

$$\therefore \ F = \overline{A}\,\overline{B} + AB + AC$$

The NAND-NAND circuit for the simplified equation is as follows.

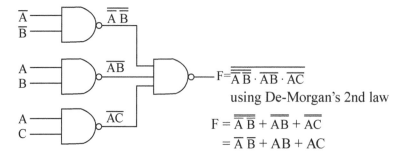

$$F = \overline{\overline{A}\,\overline{B} \cdot \overline{AB} \cdot \overline{AC}}$$

using De-Morgan's 2nd law

$$F = \overline{\overline{\overline{A}\,\overline{B}}} + \overline{\overline{AB}} + \overline{\overline{AC}}$$

$$= \overline{A}\,\overline{B} + AB + AC$$

Q.14. Minimize and implement using NOR Logic

F= Σ m (0,1,2,3,10,11,12,13,14,15)

Ans. We will minimize using Karnaugh map

F= Σ m (0,1,2,3,10,11,12,13,14,15)

$$= m_0 + m_1 + m_2 + m_3 + m_{10} + m_{11} + m_{12} + m_{13} + m_{14} + m_{15}$$

$$= \overline{A}\,\overline{B}\,\overline{C}\,\overline{D} + \overline{A}\,\overline{B}\,\overline{C}D + \overline{A}\,\overline{B}C\overline{D} + \overline{A}\,\overline{B}CD + A\overline{B}C\overline{D} + A\overline{B}CD$$

$$+ AB\overline{C}\,\overline{D} + AB\overline{C}D + ABC\overline{D} + ABCD$$

Now let us draw the Karnaugh map by placing 1 for the corresponding interms and fill the remaining places by 0.

	$\overline{C}\,\overline{D}$	$\overline{C}D$	CD	C$\overline{D}$
$\overline{A}\,\overline{B}$	1	1	1	1
$\overline{A}B$	0	0	0	0
AB	1	1	1	1
A$\overline{B}$	0	0	1	1

Fig. 1.

Now to implement the simplified equation using NOR-NOR circuit, we have to first find the POS solution for the given function. For getting POS circuit, we have to first complement each 0 and 1 on the Karnaugh map of fig 1. The complemented map is shown below:

	$\overline{C}\,\overline{D}$	$\overline{C}D$	CD	C$\overline{D}$
$\overline{A}\,\overline{B}$	0	0	0	0
$\overline{A}B$	1	1	1	1
AB	0	0	0	0
A$\overline{B}$	1	1	0	0

Now the encircled 1 s allow us to write the following sum-of-products equation with complemented output

$$\overline{Y} = \overline{A}B + A\overline{B}\,\overline{C}$$

Therefore to get the POS solution, we have to take complement on both LHS and RHS and apply De-morganis laws.

$$\therefore\ \overline{\overline{Y}} = \overline{\overline{AB} + A\ \overline{B}\ \overline{C}}$$

or $Y = \overline{\overline{AB}} . \overline{A\ \overline{B}\ \overline{C}}$ (using De-Morgan's 1st law)

$= (\overline{\overline{A}} + \overline{B})(\overline{A} + B + C)$ (using De-Morgan's 2nd law)

$\therefore Y = (A+\overline{B}) + (\overline{A}+B+C)$ is the simplified POS solution

The NOR-NOR circuit for the above equation is shown below.

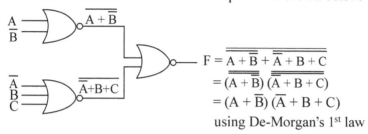

$F = \overline{\overline{A + \overline{B}} + \overline{\overline{A} + B + C}}$

$= (A + \overline{B})(\overline{A} + B + C)$

$= (A + \overline{B})(\overline{A} + B + C)$

using De-Morgan's 1st law

Q.15. Give the minimized POS solution for $Y = \pi M\ (0,3,4,5,6,7,11,15)$. Next draw the NOR-NOR circuit for the minimised equation.

Ans. $Y = \pi M\ (0, 3, 4, 5, 6, 7, 11, 15)$

To get the POS solution we will place O in the Karnaugh map for the corresponding minterms present in the given expression and place 1 in the remaining positions.

	$\overline{C}\ \overline{D}$	$\overline{C}D$	CD	$C\overline{D}$
$\overline{A}\ \overline{B}$	0	1	0	1
$\overline{A}\ B$	0	0	0	0
$A\ B$	1	1	0	1
$A\ \overline{B}$	1	1	0	1

Now we will group the 0's and each group will be written as the sum of the complement of the variable and the minimum POS will thus be the product of each such groups.

$$\therefore\ Y = (A+\overline{B})\ (\overline{C}+\overline{D})\ (A+C+D)$$

The NOR-NOR circuit for the above simplified equation is shown below

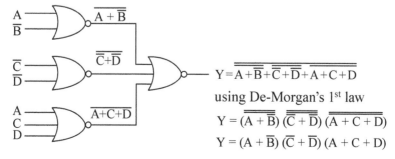

$Y = \overline{\overline{A+\overline{B}}+\overline{\overline{C}+\overline{D}}+\overline{A+C+D}}$

using De-Morgan's 1st law

$Y = (A + \overline{B})\ (\overline{C} + \overline{D})\ (A + C + D)$

$Y = (A + \overline{B})\ (\overline{C} + \overline{D})\ (A + C + D)$

Q.16. Minimize Y = F (A, B, C, D) = Σm (0, 1, 2, 4, 5, 10) +d (8, 9, 11, 12, 13, 15) using Karnaugh map.

Ans. Y = F (A, B, C, D) = Σm (0, 1, 2, 4, 5, 10) +d (8, 9, 11, 12, 13, 15)

$= m_0 + m_1 + m_2 + m_4 + m_5 + m_{10}$ +d $(m_8 + m_9 + m_{11} + m_{12} + m_{13} + m_{15})$

$= \overline{A}\,\overline{B}\,\overline{C}\,\overline{D} + \overline{A}\,\overline{B}\,\overline{C}D + \overline{A}\,\overline{B}C\overline{D} + \overline{A}\,B\overline{C}\,\overline{D} + \overline{A}\,B\overline{C}D + A\,B\overline{C}\overline{D}$
$+ d(A\,\overline{B}\,\overline{C}\,\overline{D} + A\overline{B}\,\overline{C}\,D + A\overline{B}CD + AB\,\overline{C}\,\overline{D} + A\,B\overline{C}D + ABCD)$

Now let us draw the Karnaugh map by placing 1 in the Karnaugh map for the corresponding minterms and X for the Σd minterms and the remaining places with 0.

	$\overline{C}\,\overline{D}$	$\overline{C}D$	CD	$C\overline{D}$
$\overline{A}\,\overline{B}$	1	1	0	1
$\overline{A}\,B$	1	1	0	0
$A\,B$	X	X	X	0
$A\,\overline{B}$	X	X	X	1

$\therefore$ Y = $\overline{C}$+$\overline{B}\,\overline{D}$

Q.17. Minimize Y = F (A, B, C, D, E) = Σm (0, 1, 2, 4, 5, 10) +d (8, 9, 11, 12, 13, 15, 2, 3, 24, 26, 28, 30, 31) using Karnaugh map

Ans. Y = F (A, B, C, D, E) = Σm (0, 1, 2, 4, 5, 10) +d (8, 9, 11, 12, 13, 15, 2, 3, 24, 26, 28, 30, 31).

A 5-variable map can be constructed in three dimensions by placing one 4 variable map on top of a second one. Terms in the bottom layer are numbered 0 through 15 containing A′ and corresponding terms in the top layer are numbered 16 through 31 containg A. To represent the map in two dimensions, we will divide each square in a 4-variable map by a diagonal line and place terms in the bottom layer below the line and terms in the top layer above the line. Terms in the top or bottom layer combine just like terms on a 4-variable map.

Now we will draw the Karnaugh map accordingly and place 1 for the corresponding minterms given in the expression, then do the grouping to find out the minimum SOP solution.

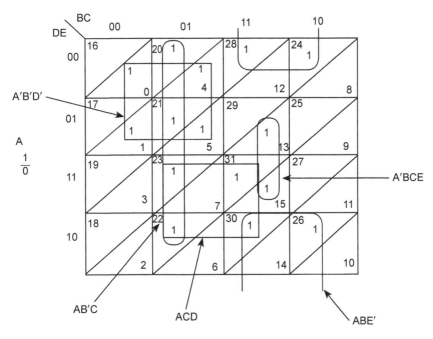

$\therefore$ Y = A'B'D' + AB'C + ACD + ABE' + A'BCE   **Ans.**

Q.18. Find the minterm expansion or standard SOP of f (a,b,c,d) = a' (b'+d) + acd'

Ans f = a'(b'+d) + acd'

    = a'b' + a'd + acd'

    = a'b'(c+c') (d+d') + a'd(b+b') (c+c') + acd' (b+b') $\because$ X + X' = 1

    = (a'b'c + a'b'c') (d+d') + (a'db + a'b'd) (c+c') + abcd' + ab'cd'

    = a'b'cd + a'b'cd' + a'b'c'd + a'b'c'd' + a'bcd + a'bc'd + a'b'cd' + a'b'c'd

        + abcd' + ab'cd'   since X + X = X

    = a'b'c'd' + a'b'c'd + a'b'cd' + a'b'cd + a'bc'd + a'bcd + abcd' + ab'cd'

    = $m_0 + m_1 + m_2 + m_3 + m_5 + m_7 + m_{14} + m_{10}$

    $\therefore$ f = Σm (0, 1, 2, 3, 5, 7, 10, 14) **Ans.**

Q.19. Show that a'c+ b'c' + ab = a'b' + bc + ac'

Ans. To solve the problem we will find the minterm expansion of each side

    L.H.S = a'c + b'c' + ab

    = a'c (b+b') + b'c'(a+a') + ab(c+c') $\because$ X + X' = 1

    = a'bc + a'b'c + ab'c' + a'b'c' + abc + abc'

    = a'b'c' + a'b'c + a'bc + ab'c' + abc' + abc

    = $m_0 + m_1 + m_3 + m_4 + m_6 + m_7$

R.H.S. = a'b' + bc + ac'

= a'b'(c + c') + bc(a + a') + ac'(b + b')

= a'b'c + a'b'c' + abc + a'bc + abc' + ab'c'

= a'b'c' + a'b'c + a'bc + ab'c' + abc' + abc

= $m_0 + m_1 + m_3 + m_4 + m_6 + m_7$

Since we find that minterm expansion for both L.H.S. and R.H.S. are the same, the equation given is valid.

Q.20. Find the minterm expansion and maxterm expansion for the function $f(a, b, c) = b + a'c$

Ans First let us find the minterm expansion for f(a, b, c) = b + a'c

= b(a + a') (c + c') + a'c(b + b') $\because$ X + X' = 1

= (ab+a'b) (c+c') + a'bc + a'b'c

= abc + abc' + a'bc + a'bc' + a'bc + a'b'c $(\because$ X + X = X)

= $m_7 + m_6 + m_3 + m_2 + m_1$

= Σm (1, 2, 3, 6, 7)

The maxterm expansion for f can be obtained by listing the decimal integers (in the range 0 to 7) which do not correspond to minterms of f

$\therefore f = \pi M$ (0, 4, 5)   **Ans.**

Q.21. Simplify the following expressions using only consensus theorem (or its dual)

(a) BC'D' + ABC' + AC'D + AB'D + A'BD'

(b) (B+C+D) (A+B+C) (A'+C+D) (B'+C'+D')

Ans. (a) BC'D' + ABC' + AC'D + AB'D + A'BD' '

= ABC' + AB'D + A'BD'   since according to consensus Theorem

XY + X'Z + YZ = XY + X'Z

(b) (B + C + D)(A + B + C) (A' + C + D) (B' + C' + D')

= (A + B + C) (A' + C + D) (B' + C' + D')

$\because$ the dual form of consensus Theorem is

(X + Y) (X' + Y) (Y + Z) = (X + Y) (X' + Z)

Q.22. Illustrate the theorems (i) A . A = A (ii) A + A = A (iii) A + 0 = A (iv) A + 1 = 1 (v) A + A' = 1.

Ans. (i) $A.A = A$ (ii) $A + A = A$

(iii) $A + 0 = A$ (iv) $A + 1 = 1$

(v) $A + A' = 1$

LONG ANSWER TYPE QUESTIONS

Q.1. Minimize the function $f(a,b,c,d) = \Sigma m$ (0, 1, 2, 5, 6, 7, 8, 9, 10, 14) using Quine-Mc Cluskey method.

Ans. Quine-Mc Cluskey method is also known as tabular method. This procedure reduces the standard S.O.P form. So if the function is not in minterm form, the minterm expansion of the given function is to be found out first. This method consists of two parts. In the first part, all the prime implicants of the function are systematically formed by combining minterms using $XY + XY' = X$ where X is a product of literals and Y is a single variable.

In the second part, we have to use a prime implicant chart to select a minimum set of prime implicants which, when ORed together are equal to the minimized form of the given function.

Now the function given is

$f(a, b, c, d) = \Sigma m$ (0, 1, 2, 5, 6, 7, 8, 9, 10, 14)

First the binary minterms are sorted into groups according to the number of 1's in each term. Next, terms in adjacent groups must be compared and eliminate the variable and the resulting terms are listed in column II. Further in the same way we will compare and eliminate variable and resulting terms to be listed in column III. It is to be noted that terms in the first group in Column II need only be compared with terms in the second group which have dashes in the same places. Again, the terms which have been combined are checked off. If there are duplicate terms in column III, we have to delete the duplicate terms. This way when no further combination is possible, the process terminates. Accordingly the table for the given function is:

	Column I	Column II	Column III

group 0   0 0000✓   0, 1 000–✓   0,1,8,9   –00–
             0, 2 00–0✓   0,2,8,10   –0–0

group 1   1 0001✓   0, 8 –000✓   0,8,1,9   ~~–00–~~
             2 0010✓   1, 5 0–01   0,8,2,10   ~~–0–0~~
             8 1000✓   1, 9 –001✓   2,6,10,14 ––10

group 2   5 0011✓   2,6 0–10✓   ~~(2,10,6,14)~~ ~~––10~~
             6 0110✓   2,10 –010✓
             9 1001✓   8,9 100–✓
             10 1010✓   8,10 10–0✓

group 3   7 0111✓   5,7 01–1
             14 1110✓   6,7 011–
                          6,14 –110✓
                          10,14 1–10✓

The terms which have not been checked off because they cannot be combined with other terms are called prime implicants. Since every minterm has been inclueded in at least one of the prime implicants, the function is equal to the sum of its prime implicants.

$$\therefore f = a'c'd + a'bd + a'bc + b'c' + b'd' + cd'$$

(1,5)   (5,7)   (6,7)   (0,1,8,9) (0,2,8,10) (2,6,10,14)

In the above expression, each term has a minimum number of literals, but the number of terms is not minimum.

So, according to second part of Quine McCluskey method, we will now draw the prime implicant chart. The minterms of the function are listed across the top of the chart and the prime implicants are listed down the side. A prime implicant is equal to a sum of minterms and the prime implicant is said to cover these minterms. If a prime implicant covers a given minterm, an X is placed at the intersection of the corresponding row and column

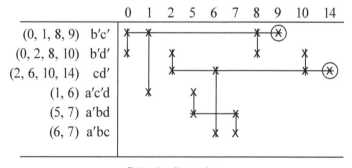

Prime implicant chart

The encircled prime implicants are know as essential prime implicants must be included in the minimum SOP solution. Now start with one essential prime imlicant and cross out corresponding row. Then the columns which correspond to all minterms covered by that prime implicant should also be crossed out. After this a minimum set of prime implicants must now be chosen to cover the remaining columns. Here we find a′bd covers the remaining two columns, so we will select it. Hence the resulting minimum SOP is

f = b′c′ + cd′ + a′bd   **Ans.**

Q.2. Minimize the function f (a,b,c,d) = Σm(2,3,7,9,11,13) + Σd (1, 10, 15) using Quine McCluskey method

Ans. f(a, b, c, d) = Σm (2, 3, 7, 9, 11, 13) + Σd (1, 10, 15)

The don't cares are treated like required minterms when finding the prime implicants.

First the binary minterms are sorted into groups according to the number of 1's in each term. Next, terms in adjacent group must be compared and eliminate the variable using XY + XY′ = X where X is a minterm and Y is a single variable. The resulting terms are listed in Column II. Same way we will further proceed in Column II and the resulting terms to be listed in Column III. Also the terms which have been combined are checked off. The duplicate terms in column III are also to be deleted. Accordingly the table for the given function is as follows:

	Column I		Column II		Column III	
group 1	1	0001✓	(1,3)	00–1✓	(1,3,9,11)	–0–1
	2	0010✓	(1,9)	–001✓	(1,9,3,11)	–0–1
group 2	3	0011✓	(2,3)	001–✓	(2,3,10,11)	–01–
	9	1001✓	(2,10)	–010✓	(2,10,3,11)	–01–
	10	1010✓	(3,7)	0–11✓	(3,7,11,15)	– –11
group 3	7	0111✓	(3,11)	–011✓	(3,11,7,15)	– –11
	11	1011✓	(9,11)	10–1✓	(9,11,13,15)	1– –1
	13	1101✓	(9,13)	1–01✓	(9,13,11,15)	1– –1
group 4	15	1111✓	(10,11)	101–✓		
			(7,15)	–111✓		
			(11,15)	1–11✓		
			(13,15)	11–1✓		

The terms which have not been checked off because they cannot be combined with other terms are called prime implicants. The function is now equal to the sum of its prime implicants.

$$\therefore f = \quad b'd \quad + \quad b'c \quad + \quad cd \quad + \quad ad$$

(1,3,9,11)   (2,3,10,11)   (3,7,11,15)   (9,11,13,15)

In the above expression, each term has a minimum number of literals, but the number of terms is not minimum.

To get the minimum solution we will now draw the prime chart. The don't care columns are omitted when forming the prime implicant chart.

The minterms of the function are listed across the top of the chart and the prime implicants are listed down the side. If a prime implicant covers a given minterm, an X is placed at the intersection of the corresponding row and column

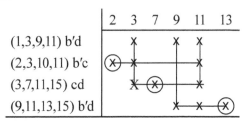

Prime Implicant chart

The encircled prime implicants are known as essential prime implicant as they cover only one minterm and these should be included in the minimum sum-of-products solution.

Therefore starting with the essential prime implicants and crossing out the prime implicants of the corresponding rows and thereby columns, the resulting minimum S.O.P solution is

$f = b'c + cd + ad$ **Ans.**

5

Binary Arithmetic

FACTS THAT MATTER

1. **The four Basic arithmetic operations are**
   - (*i*) Addition
   - (*ii*) Subtraction
   - (*iii*) Multiplication
   - (*iv*) Division.
2. **The basic rules for binary addition are**
   - (*i*) $0+0 = 0$
   - (*ii*) $0+1 = 1$
   - (*iii*) $1+0 = 1$
   - (*iv*) $1+1 = 0$ with carry 1.
3. **The basic rules for binary subtraction are**
   - (*i*) $0-0 = 0$
   - (*ii*) $1-0 = 1$
   - (*iii*) $1-1 = 0$
   - (*iv*) $10-1 = 1$

     Note : In the fourth case you have to borrow from the next higher column while subtracting.
4. **L.S.B:** Least significant bit.
5. **M.S.B:** Most significant bit.
6. **Ripple carry:** Carry that ripples from one stage to other in serial addition.

7. **Serial addition:** A method of binary addition where carry sequentially propagates from one stage to next stage.

8. **Magnitude:** The absolute or unsigned value of a number

9. **Unsigned binary number:** Here all of the bits in a binary number are used to represent the magnitude of the corresponding decimal number.

10. **Sign-magnitude number:** If the number contains a sign bit followed by magnitude bits, it is known as sign-magnitude number.

11. The sign bit $0 \rightarrow$ means positive binary number

$1 \rightarrow$ means negative binary number.

12. **Arithemetic logic unit:** A device than can perform both arithmetic and logic function based on select inputs

13. **Microprocessor:** A digital IC that combines the arithmetic and control sections of a computer.

14. Range of unsigned Binary number;

e.g. The smallest 8 bit number is 00000000

i.e. (OOH). The largest 8 bit number is 1111 1111

i.e. (FFH)

This is equivalent to a decimal 0 to 255.

15. **Limit:** The result of addition must be within the range of that magnitude. e.g. In adding 8 bit unsigned number sum should fall in the range of 0 to 255. If any magnitude is greater than 255, you should use 16-bit arithmetic.

16. **Overflow:** An unwanted carry that produces an answer outside the valid range of the numbers being represented. For e.g. In 8-bit arithmetic, if the sum of two unsigned numbers is greater than 255, it causes an overflow, a carry into the ninth column.

17. **Carry flag:** Most microprocessors have a logic circuit called carry flag which detects the overflow and warns you that the answer is invalid.

18. **1's Complement:** The 1's complement of a binary number is obtained by converting 1s to 0s and 0s to 1s.

19. **2's Complement:** The 2's complement of a binary number can be derived by adding 1 to the 1's complement. The advantage of 2's complement over 1's complement is that of not requiring an end-around carry during addition.

20. **Half adder:** Half Adder is a logic circuit with two inputs and two outputs which adds 2 bits at a time, producing a sum and a carry output.

21. **Full adder:** Full Adder is a logic circuit with three inputs and two outputs which adds 3 bits at a time, giving a sum and carry output.

22. **Half Subtractor:** Half Subtractor is a logic circuit with 2 inputs and 2 outputs which subtracts one bit from the other producing a difference and a borrow output.

23. **Full Subtractor:** A Full subtractor performs the subtraction involving three bits *i.e.*, it taken into account the minuend bit, subtrahend bit and the borrow from the previous stage giving 2 outputs as difference and borrow to be carried to the next stage.

OBJECTIVE TYPE QUESTIONS

1. What kind of number is 179 FH?
2. What is the meaning of 111_2? and 111_{10}?
3. What is the carry flag in a microprocessor?
4. What is the largest decimal number that can be represented with an 8-bit unsigned binary number ?
5. What is the decimal number range that can be represented with an 8-bit sign-magnitude binary number ?
6. In sigh-magnitude form, what is the decimal value of 10001101? of 00001101?
7. What is the 1's compliment representation of 11010110?
8. What is the 2's compliment representation of 11010110?
9. What is the standard method for doing binary arithmetic in nearly all microprocessors?
10. How is 2's complement representation used to perform subtraction?
11. What are the inputs and outputs of a half-adder?
12. What are the inputs and outputs of a full-adder?
13. What is an ALU?
14. Which gate can easily implement the SUM output of a full-adder?
15. How one can do division of binary numbers?

Answers

1. This is the hexadecimal number 179F.
2. Binary 111; decimal 111.
3. It is used to indicate an overflow.
4. 255.
5. −127 to +127.
6. −13; + 13.

7. 00101001.

8. 00101010.

9. 2's complement.

10. Take the complement of the subtrahend and add it to the minuend.

11. Inputs : A and B; outputs: sum and carry.

12. Inputs: A,B and CARRY IN ; outputs: Sum and Carry.

13. ALU is short form of Arithmetic Logic Unit, a digital hardware that can perform both arithmetic and logic operations.

14. Exclusive –OR gate.

15. It is realized only by repeatedly subtracting one number from the other.

SHORT ANSWER TYPE QUESTIONS

Q.1. Give the sum in each of the following:

(a) $3_8 + 7_8 = ?$ (b) $5_8 + 6_8 = ?$

(c) $4_{16} + C_{16} = ?$ (d) $8_{16} + F_{16} = ?$

Ans. (a) $3_8 = 011_2$, $7_8 = 111_2$

$$011$$
$$+ \; 111$$
$$\underline{001010}$$

$\therefore 3_8 + 7_8 = 12_8$ **Ans.**

$$\;\;1\;\;\;\;2$$

(b) $5_8 = 101_2$, $6_8 = 110_2$

$$101$$
$$+110$$
$$\underline{001011}$$

$\therefore 5_8 + 6_8 = 13_8$ **Ans.**

$$\;\;1\;\;\;\;3$$

(c) $4_{16} = 0100_2$, $C_{16} = 1100_2$

$$0100$$
$$+1100$$
$$\underline{00010000}$$

$\therefore 4_{16} + C_{16} = 10_{16}$ **Ans.**

$$\;\;1\;\;\;\;0$$

(d) $8_{16} = 1000_2$, $F_{16} + 1111_2$

$$1000$$
$$+1111$$
$$\underline{00010111}$$

$\therefore 8_{16} + F_{16} + 17_{16}$ **Ans.**

$$\;\;1\;\;\;\;7$$

Q.2. Work out each of these binary sums:

    (a) $00010100 + 00101001$

    (b) $0001100011110110 + 0000111100001000$

Ans. (a) 00010100

      $+\underline{00101001}$

        00111101 **Ans.**

    (b) 0001100011110110

       $+\underline{0000111100001000}$

        0010011111111110 **Ans.**

Q.3. Show the binary addition of 750_{10} and 538_{10} using 16 bit numbers.

Ans. First let us convert the given numbers to their equivalent binary numbers.

2	750			2	538	
2	375	0		2	269	0
2	187	1		2	134	1
2	93	1		2	67	0
2	46	1		2	33	1
2	23	0		2	16	1
2	11	1		2	8	0
2	5	1		2	4	0
2	2	1		2	2	0
	1	0			1	0
		1				1

$\therefore 750_{10} = 1011101110_2$ $\therefore 538_{10} = 1000011010_2$

Now for the 16 bit binary addtion of the above numbers

    0000 0010 1110 1110

    <u>0000 0010 00011010</u>

    0000 0101 00001000$_2$ **Ans.**

Q.4. Subtract the following

      (a) $01001111 - 00000101$

      (b) $47_{10} - 23_{10}$ in binary form

Ans. (a) 01001111

      $-\underline{00000101}$

       01001010 **Ans.**

(b) $47_{10} = 101111_2$, $23_{10} = 10111$

    0101111

    $-\underline{010111}$

     011000 **Ans.**

Q.5. Express each of the following in 8-bit sign-magnitude form:

    (a) + 23   (b) –107

Ans. (a) 32 16 8 4 2 1 → binary weights

            1 0 1 1 1 → 23_{10}

    ∴ In 8 bit sign - magnitude form

        + 23 = 00010111 **Ans.**

    (b) 64 32 16 8 4 2 1 → binary weights

          1 1 0 1 0 1 1 → 107_{10}

    ∴ In 8 bit sign – magnitude form

        –107 = 11101011 **Ans.**

Q.6. Convert each of the sign-magnitude numbers into decimal equivalents

    (a) 00110110   (b) 11111000

Ans. (a) 32 16 8 4 2 1 → binary weights

           1 1 0 1 1 0 → 54_{10}

    ∴ 00110110 = + 54_{10} **Ans.**

    (b) 64 32 16 8 4 2 1 → binary weights

          1 1 1 1 0 0 0 → 120_{10}

    ∴ 1 1 1 1 1 0 0 0 = -120_{10} **Ans.**

Q.7. Express the 1's complement of each of the following in hexadecimal notation:

    (a) 23H   (b) FDH

Ans. (a)   2   3

          ↓   ↓

       0010   0011

    Now 1's complement of $00100011_2 = 11011100_2$

    In hexadecimal notation 11011100_2 = DCH. **Ans.**

    (b) F   D

        ↓   ↓

      1111   1101

    Now 1's complement of $11111101_2 = 0000\ 0010_2$

    In hexadecimal notation $0000\ 0010_2$ = 02 H. **Ans.**

Q.8. What is the 2's complement of each of these:

        (a) 0000 1111   (b) 1011 1110

Ans. (a) 2's complement = 1's complement +1

    ∴ 2's complement of 0000 1111 = 1111 0000

$$\frac{\qquad\qquad +1}{1111\ 0001}\ \textbf{Ans.}$$

(b) 2's complement of 1011 1110 = 0100 0001

$$\frac{+1}{0100\ 0010}\quad\textbf{Ans.}$$

Q.9. Convert each of the following to 2's complement representation

      (a) +78   (b) –23   (c) –90   (d) –121

Ans.   64 32 16 8 4 2 1

     1 0 0 1 1 1 0 → 78

∴ 2's complement representation of +78

   = 01001110

which is same as 8 bit binary equivalent of 78

(b) –23

Here first we have to convert the magnitude to binary form (8 bit)

128 64 32 16 8 4 2 1

  0 0 0 1 0 1 1 1

Now take the 2's complement to obtain the negative value which is

   1 1 1 0 1 0 0 1 **Ans.**

(c) –90

   First we have to convert the magnitude to binary form (8 bit)

   128 64 32 16 8 4 2 1

    0 1 0 1 1 0 1 0 → 90

Now take the 2's complement to obtain the negative value which is

   1010 0110   **Ans.**

(*d*) –121

   First we have to convert the magnitude to binary form (8 bit)

   128 64 32 16 8 4 2 1

    0 1 1 1 1 0 0 1 → 121

   Now take the 2's complement to obtain the negative value which is

   1 0 0 0 0 1 1 1 **Ans.**

Q.10. Perform the follwoing arithmetic operations:

      (a) AC 2B5 + DEFA4

      (b) F25CF – 9BAC4

Ans. (a) A   C   2   B   5   D   E   F   A   4

       1010 1100 0010 1011 0101   1101 1110 1111 1010 0100

           1010 1100 0010 1011 0101

        + 1101 1110 1111 1010 0100

         11000 1011 0010 0101 1001   **Ans.**

(b) F   2   5   C   F   9   B   A   C   4
    1111 0010 0101 1100 1111   1001 1011 1010 1100 0100
          1111 0010 0101 1100 1111
        − 1001 1011 1010 1100 0100
          0101 1110 1011 0000 1011   **Ans.**

Q.11. Perform $25_{10} - 16_{10}$ using 1's complement and 2's complement method.

Ans. Subtraction by 1's complement method:

$\qquad$ Minuend $= 25_{10} = 11001_2$

$\qquad$ Subrahend $= 16_{10} = 10000_2$

First we have to take 1's complement of subrahend.

$\qquad\qquad = 01111$

Next add 1's complement of subrahend to minuend

$\quad$ *i.e.* $\qquad\qquad$ 11001

$\qquad\qquad\qquad$ 01111

end around carry $\rightarrow$ 101000 $\rightarrow$ Remainder

∵ there is an end-around carry, we have to add it to the remainder, which gives the final answer.

$\qquad\qquad$ 01000

$\qquad\qquad$ + 1

$\qquad\qquad$ 01001   **Ans.**

Subtraction by 2's complement method:

First we have to take 2's complement of subrahend $= 01111 + 1 = 10000$

Next add 2's complement of subrahend to Minuend *i.e.*

$\qquad\qquad$ 11001

$\qquad\qquad$ + 10000

end around carry $\rightarrow$ 101001 $\rightarrow$ Remainder

Since there is end-around carry, discard the carry, remainder is the final answer.

∴ 01001   **Ans.**

Q.12. Subtract 27_{10} from, 17_{10} using 1's complement and 2's complement method.

Ans. Subtraction by 1's complement method;

$\qquad$ Minuend $= 17_{10} = 10001_2$

$\qquad$ Subrahend $= 27_{10} = 11011_2$

First we have to take 1's complement of subrahend $= 00100$

Next add 1's complement of subrahend to Minuend

$\quad$ *i.e.* $\qquad\qquad$ 10001

00100

10101   → 1st result

Since there is no end-around carry, we have to take 1's complement of 1st result, give a –ve sign to get the final Ans.

∴   –01010 **Ans.**

Subtraction by 2's complement method;

First we have to take 2's complement of subtrahend = 1's complement +1

$$= 00100 + 1 = 00101$$

Now add this to the minuend

i.e.   10001

00101

10110   → 1st result

Since there is no end-around carry, we have to take 2's complement of the 1st result give a –ve sign to get the final result.

∴   01001 + 1 = 01010

Hence   –01010   **Ans.**

Q.13. Add +83 and +16

Ans. 64 32 16 8 4 2 1 → binary weights 32 16 8 4 2 1 → binary weights

1 0 1 0 0 1 1 → 83   1 0 0 0 0 → 16

∴ + 83 = 0101 0011

+ 16 = 0001 0000

+ 99   0110 0011

∴ Binary answer is 0110 0011 which is equal to 63 H

Now 99 = 16 $\lfloor$99

6   3 ↑

6 $\vert$

= 63 H

This agrees the decimal sum

Hence 0110 0011 **Ans.**

Q.14. Add +7 and –4

Ans. +7 = 0111

– 4 = 2's complement of 0100 = 1011 + 1 = 1100

∴   +7 =   0111

+ (–4) = + 1100

+ 3   1 0011

Since there is end-around carry, discard the carry

∴   0011 **Ans.**

Q.15. Add +37 and –115

Ans. 32 16 8 4 2 1   → binary weights   → 64 32 16 8 4 2 1

       1 0 0 1 0 1 → 37   1 1   1 0 0 1 1 → 115

∴ + 37 = 0010 0101

  –115 = 2's complement of 0111 0011 = 1000 1101

∴ + 37 =   0010 0101

+(–115) = + 1000 1101

   –78   1011 0010

To verify the answer +78 = 0100 1110

  –78 = 2's complement of 0100 1110

            = 1011 0010

∴ 1011 0010 **Ans.**

Q.16. Add –20 and –18

Ans. 16 8 4 2 1 → binary weight → 16 8 4 2 1

       1 0 1 0 0 → 20   1 0 0 1 0 → 18

∴ + 20 = 0001 0100

∴ – 20 = 2's complement of 0001 0100

                        = 1110 1011 + 1

                        = 1110 1100

+ 18 = 0001 0010

∴ –18 = 2's complement of 0001 0010

                        = 1110   1110

∴ –20 =   1 1 1 0 1 1 0 0

+ (–18) = + 1 1 1 0 1 1 1 0

   –38   1 1 1 0 1 1 0 1 0

Now discard the end around carry, the remainder is the answer 1101 1010

To verify the answers.

+38 = 0010 0110

∴ –38 = 2's complement of 0010 0110 = 1101 1010

Hence 1101 1010 **Ans.**

Q.17. Subtract +16 from +83

       + 83 = 0101 0011

       + 16 = 0001 0000

To subtract +16, we have to first take 2's complement of 00010000 to get

   –16 = 1 1 1 0 1 1 1 1 + 1 = 11110000

Next add this to +83

∴ + 83   0 1 0 1 0 0 1 1
+(–16)   1 1 1 1 0 0 0 0
    67   1 0 1 0 0 0 0 1 1

Since there is en-around carry, discard the carry ∴ 0100 0011 **Ans.**

Q.18. Subtract –27 from + 68

Ans. + 68 = 0100 0100

     + 27 = 0001 1011

     ∴ –27 = 2's complement of 00011011 = 11100101

Since we have to subtract –27 from +68

∴ we take 2's complement of (–27) = 00011011

∴   + 68 =   +68   0100 0100
      –(–27) =   +27   + 0001 1011
                +95   0101 1111   **Ans.**

Q.19. Subtract –7 from 5

Ans. + 5 = 0101

     + 7 = 0111

     ∴ –7 = 2's complement of 0111 = 1001

Now +5 – (–7) = +5 +7 = + 12

Now 2's complement of –7 = +7 = 0111

∴   0101
            + 0111
            1100   **Ans.**

Q.20. Subtract –78 from –43

Ans +43 = 00101011

     ∴ –43 = 2's complement of 00101011

             = 1101 0101

     + 78 = 0100 1110

     ∴ –78 = 2's complement of 01001110

                 = 10110010

To subtract –78, first take 2's complement of –78 to get

            + 78 = 0100 1110

Next add to obtain

     –43   1101 0101
     +78   0100 1110
      35   1 0010 0011

Discard the end-around carry

∴ 0010 0011 **Ans.**

Q.21. Multiply

(*a*) 1001 by 101,   (*b*) F5 × A

Ans (a)   1001
                      101
                     ‾‾‾‾
                     1001
                     0000
                    1001
                   ‾‾‾‾‾‾
                   101101 **Ans.**

(b)   F   5   A
        ↓   ↓   ↓
      1111 0101 1010

               1111 0101
                    1010
              ‾‾‾‾‾‾‾‾‾‾‾
             : : : :
             : 00000000
             : 11110101
             : 00000000
             11110101
            ‾‾‾‾‾‾‾‾‾‾‾‾‾
            100110010010   **Ans.**

Q.22. Divide

(a) 110_2 by 10_2 (b) 1111001_2 by 1001_2

Ans. (a)   11
                      ‾‾‾‾‾‾‾
                  10)110
                       10
                      ‾‾‾
                       10
                       10
                      ‾‾‾
                        0

∴ Quotient = 11, Remainder = 0 **Ans.**

                              1101
                          ‾‾‾‾‾‾‾‾‾
(b)   1001)1111001
                        1001
                       ‾‾‾‾‾
                        1100
                        1001
                       ‾‾‾‾‾
                        001101
                          1001
                         ‾‾‾‾‾
                          100   ∴ Quotient = 1101, Remainder = 100 **Ans.**

Q.23. Draw a half adder using NAND gates only.

Ans. We know for a half-adder

$$SUM = \overline{A}B + A\overline{B}$$

$$CARRY = AB \qquad \text{Where A and B are two inputs}$$

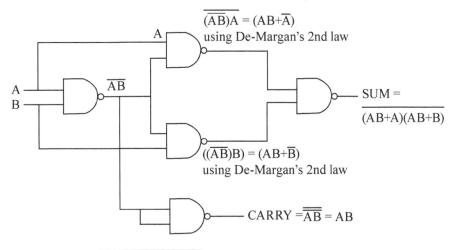

$$\overline{(\overline{AB})A} = (AB+\overline{A})$$
using De-Margan's 2nd law

$$\overline{((\overline{AB})B)} = (AB+\overline{B})$$
using De-Margan's 2nd law

$$SUM = \overline{(AB+A)(AB+B)}$$

$$CARRY = \overline{\overline{AB}} = AB$$

$$SUM = \overline{(AB + \overline{B})(AB + \overline{B})}$$

$$= \qquad \overline{(AB + \overline{A})} + \overline{(AB + \overline{B})} \qquad \text{using De-Morgan's 2nd law}$$

$$= \qquad \overline{(AB)}\,\overline{\overline{A}} + \overline{(AB)}\,\overline{\overline{B}} \qquad \text{using De-Morgan's 1st law}$$

$$= \qquad (\overline{A} + \overline{B})\,A + (\overline{A} + \overline{B})\,B \qquad \text{using De-Morgan's 2nd law}$$

$$= \qquad A\overline{A} + A\overline{B} + \overline{A}\,B + B\overline{B}$$

$$= \qquad 0 + A\overline{B} + \overline{A}B + 0$$

$$= \qquad \overline{A}B + A\overline{B}$$

Hence the half-adder is realised by NAND gates only as shown above.

LONG ANSWER TYPE QUESTIONS:

Q.1. What is a half-adder? Derive its truth table

Ans. Half-adder is a logic circuit with two inputs and two outputs which adds 2 bits at a time, producing a sum and a carry outputs.

The basic circuit of half adder is shown below

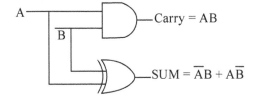

Now to derive its truth table, let us consider all the possible cases

Case 1: if A = 0, B = 0

then carry = AB = 00 = 0

$$SUM = \overline{A}B + A\overline{B} = \overline{0}\,0 + 0\,\overline{0} = 1.0 + 0.1$$

$$= 0 + 0 = 0$$

Case II:

If A = 0, B = 1

Then carry = AB = 0.1 = 0

$$SUM = \overline{A}B + A\overline{B} = \overline{0}.1 + 0.\overline{1} = 1.1 + 0.0$$

$$= 1 + 0 = 1$$

Case III:

If A = 1, B = 0

then carry = AB = 1.0 = 0

$$SUM = \overline{A}B + A\overline{B} = \overline{1}.0 + 1.\overline{0} = 0.0 + 1.1$$

$$= 0 + 1 = 1$$

Case IV:

If A = 1, B = 1

then carry = AB = 1.1 = 1

$$SUM = \overline{A}B + A\overline{B} = \overline{1}.1 + 1.\overline{1} = 0.1 + 1.0 = 0 + 0$$

$$= 0$$

Since there are two inputs A and B, therefore the above are the four possible cases. Hence the truth table of Half-Adder will be as follows.

Inputs		Outputs	
A	B	Carry	SUM
0	0	0	0
0	1	0	1
0	0	0	1
1	1	1	0

Truth table of half-adder

Q.2. What is a half-subtractor? Derive its truth table

Ans. Half subtractor is a logic circuit with 2 inputs and 2 outputs which subtracts one bit from the other producing a difference and a borrow output.

THe basic circuit of half subtractor is shown below:

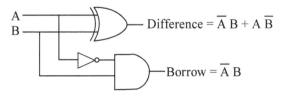

Now to derive its truth table, let us consider all the possible cases. Since there are two inputs, there will be four possible cases.

Case 1:

If $A = 0$, $B = 0$

Then difference $= \overline{A}B + A\overline{B} = \overline{0}.0 + 0.\overline{0} = 1.0 + 0.1$

$$= 0+0 = 0$$

Borrow $= \overline{A}B = \overline{0}.0 = 1.0 = 0$

Case II:

If $A = 0$, $B=1$

Then difference $= \overline{A}B + A\overline{B} = \overline{0}.1 + 0.\overline{1} = 1.1 + 0.0$

$$= 1 + 0 = 1$$

Borrow $= \overline{A}B = \overline{0}.1 = 1.1 = 1$

Case III:

If $A = 1$, $B = 0$

Then difference $= \overline{A}B + A\overline{B} = \overline{1}.0 + 1.\overline{0} = 0.0 = 1.1$

$$= 0 + 1 = 1$$

Borrow $= \overline{A}B = \overline{1}.0 = 0.0 = 0$

Case IV

If $A = 1$, $B=1$

Then difference $= \overline{A}B + A\overline{B} = \overline{1}.1 + 1.\overline{1} = 0.1 + 1.0$

$$= 0 + 0 = 0$$

Borrow $= \overline{A}B = \overline{1}.1 = 0.1 = 0$

Hence the truth table of Half-subtractor will be as follows:

Inputs		Outputs	
A	B	Difference	Borrow
0	0	0	0
0	1	1	1
1	0	1	0
1	1	0	0

Truth table of half-subtractor

Q.3. What is Full adder? Give the truth table of a Full-adder and design its logic circuit using Karnaugh map.

Ans. Full-adder is a logic circuit with three inputs and two outputs which adds 3 bits at a time giving a sum and a carry output. The third bit which it adds is the carry from the lower column.

The truth table of a Full-Adder is shown below:

Inputs			Outputs	
A	B	C	CARRY	SUM
0	0	0	0	0
0	0	1	0	1
0	1	0	0	1
0	1	1	1	0
1	0	0	0	1
1	0	1	1	0
1	1	0	1	0
1	1	1	1	1

Now from the truth table let us draw the Karnaugh map for CARRY

C\AB	00	01	11	10
0	0	0	1	0
1	0	1	1	1

$\therefore$ CARRY = AB + BC + AC

Now let us draw the Karnaugh map for SUM

C\AB	00	01	11	10
0	0	1	0	1
1	1	0	1	0

$\therefore$ we find that output is 1 for odd no.of one's

$\therefore$ SUM = A $\oplus$ B $\oplus$ C

Hence the design of the logic circuit for Full-Adder can be as follows:-

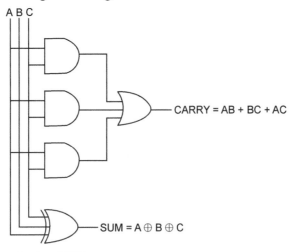

CARRY = AB + BC + AC

SUM = A $\oplus$ B $\oplus$ C

FULL - ADDER

Q.4. Give the truth table of a full subtractor and design its logic circuit using Karnaugh map.

Ans. A full subtractor performs the subtraction involving three bits i.e. it takes into account the minuend bit, subtrahend bit and the borrow from the previous stage. Let the three inputs be X, Y and Co corresponding to minuend bit, subtrahend bit and borrow from the subtraction process of previous stage. The outputs are the difference D and the borrow to be carried to the next stage.

To construct the truth table we should subtract Y+Co from X and look for the Difference D i.e. X – (Y+Co) = D. If X>(Y+Co) then Borrow B = 0. If X < (Y+ Co), then B =1 since we can subtract only after borrowing 1 from the next stage.

Hence the truth table for Full-subtractor can be constructed as shown below:

X	Y	Co	B	D
0	0	0	0	0
0	0	1	1	1
0	1	0	1	1
0	1	1	1	0
1	0	0	0	1
1	0	1	0	0
1	1	0	0	0
1	1	1	1	1

Truth table of Full Subtractor

Now from the truth table, let us draw the Karnaugh map for Borrow

X\YCo	00	01	11	10
0	0	1	1	1
1	0	0	1	0

$$\therefore B = YCo + \overline{X}Y + \overline{X}Co$$

Now let us draw the Karnaugh map for Difference

X\YCo	00	01	11	10
0	0	1	0	1
1	1	0	1	0

$$\therefore D = \overline{X}\,\overline{Y}\,Co + \overline{X}\,Y\,\overline{Co} + X\,\overline{Y}\,\overline{Co} + XYCo$$
$$= X \oplus Y \oplus Co$$

Hence the design of the logic circuit for Full subtractor can be as shown below:

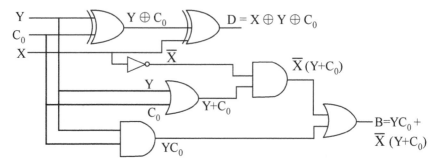

Full Subtractor

Q.5. Explain how Binary Adder-Subtractor circuit can do addition as well as subtraction.

Ans. We can connect full adders as shown below to form an Adder-subtractor circuit which can add or subtract binary numbers.

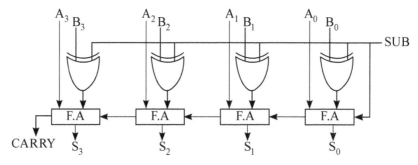

The circuit is laid out from right to left, similar to the way we add binary numbers. The least significant column is on the right and the most significant column is on the left. The boxes labeled F.A are full-adders. The carry-out from each full-adder is the carry-in to the next-higher full-adder.

In the figure shown the binary numbers taken are $A_3 A_2 A_1 A_0$ and $B_3 B_2 B_1 B_0$

For Addition, the SUB signal is deliberately kept in the low state i.e. 0. Therefore the binary number $B_3B_2B_1B_0$ passes through the Exclusive-OR gate with no change (as we know from the truth table of X–OR gate).

The full adders then produce the correct output sum by adding the bits in each column, passing carries to the next higher column. For e.g., starting at the L.S.B, the full-adder adds A_0, B_0 and SUB. This produces a SUM of S_0 and a carry out to the next higher full-adder. The next-higher full-adder then adds, A_1, B_1, and the carry-in to produce S_1 and a carry out. A similar addition occurs for each of the remaining full-adders and the correct sum appears at the output lines.

Let the two binary numbers be

$A_3 A_2 A_1 A_0 = 1\ 0\ 1\ 1$ and $B_3 B_2 B_1 B_0 = 0110$. Since SUB = 0, the first full-adder performs addition as $0 + 1 + 0 = 1$ with a carry of 0 $\therefore S_0 = 1$.

The carry-out of the first full-adder is the carry-in of the second full-adder.

$\therefore$ Second full adder performs addition as $0 + 1 + 1 = 0$ with a carry of 1 $\therefore S_1 = 0$. Similarly the third full-adder will perform addition as $1+0+1=0$ with a carry of 1 $\therefore S_2 = 0$.

The last full-adder will perform addition as $1 + 1 + 0 = 0$ with a final carry of 1

$\therefore S_3 = 0$

$\therefore$ The output will appear as $10001 = 17_{10}$

Now Data taken were $1011 = 11_{10}$ and $0110 = 6_{10}$

$\therefore 11_{10} + 6_{10} = 17_{10}$

For Subtraction

The SUB signal is now deliberately kept in the high state i.e. 1. Therefore the Exclusive-OR gate produces the 1's complement of B_3---B_0. Now assuming that A_3---$A_0 = 0$, since SUB is the Carry-IN to the first full-adder, the circuit produces the 2's complement of B_3---B_0 because 1 is being added to 1's complement of B_3---B_0. When A_3---A_0 does not equal zero, the effect is equivalent to adding A_3---A_0 and the 2's complement of B_3---B_0.

Now let us take the same data as above i.e. $A_3A_2A_1A_0 = 1011$, and $B_3B_2B_1B_0 = 0110$. 1's complement of $0110 = 1001$

Since SUB = 1, the first F.A performs addition as $1 + 1 + 1 = 1$ with a carry of 1.

$\therefore\quad S_0 = 1$

Second FA performs addition as $1 + 1 + 0 = 0$ with a carry of 1

$\therefore\quad S_1 = 0$

Third F.A. performs addition as $1 + 0 + 0 = 1$ with a carry of 0.

$\therefore\quad S_2 = 1$

The last F.A. performs addition as $0 + 1 + 1 = 0$ with a carry of 1

$\therefore\quad S_3 = 0$

$\therefore$ The output with appear as $1\ 0\ 1\ 0\ 1$

Since in a 2's complement addition the end-around carry is discarded.

$\therefore$ The final answer is $0\ 1\ 0\ 1$ which is equal to 5_{10}

Now Data taken were $1\ 0\ 1\ 1 = 11_{10}$ and $0110 = 6_{10}$

$\therefore\quad 11_{10} - 6_{10} = 5_{10}$

Hence we find that if the SUB input signal is low or 0, the circuit act as Adder circuit and if SUB is kept high or 1, the same circuit act as subtractor circuit.

Codes and Parity

FACTS THAT MATTER

1. Computers and other digital circuits are required to handle data which may be numeric, alphabetic or special characters. Since computer and digital circuits work with binary numbers, therefore a code is needed to convert the given data into binary numbers. This process is known as encoding.

2. There are various types of codes used in digital system. They are categorized as:

   (*i*) Weighted binary codes

   (*ii*) Non-weighted codes

   (*iii*) Alphanumeric codes

   (*iv*) Error detection codes

3. **Weighted Binary Codes:** In weighted codes for each position (or bit), there is specific weight attached. Several weighted codes are possible *e.g.*, 8421, 2421, 4221, 5421, 6311, 7421 etc. All such codes are also known as binary coded decimal or BCD code.

   The most common weighted code is 8421. Since 8421 BCD is the most natural amongst the other possible code, it is often referred to as BCD without describing it. So by default BCD means 8421. In any BCD, each decimal digit is expressed by its 4 bit binary equivalent according to the BCD code being used for e.g. decimal 567 can be encoded as follows in various 4 bit codes

Decimal number	→	5	6	7
8421 code	→	0101	0110	0111
5421 code	→	1000	1001	1010
6311 code	→	0111	1000	1001

4. **Non-weighted codes:** In non-weighted codes there is no specific weight attached for each position (or bit).

There are two type of non-weighted code:

(*i*) Excess-3 code

(*ii*) Gray code

5. **Excess-3 code:** This is an important non-weighted BCD code. For converting decimal number in excess-3 code (also called XS-3), we add 3 to each decimal digit before converting it to equivalent binary in 4 bits. *i.e.* each 4 bit group in XS-3 code is equal to a specific decimal digit.

Table below shows 8421 BCD and Excess-3 code equivalent to some decimal numbers

Decimal number	8421 BCD code	Excess-3 code
0	0000	0011
1	0001	0100
2	0010	0101
3	0011	0110
4	0100	0111
5	0101	1000
6	0110	1001
7	0111	1010
8	1000	1011
9	1001	1100
10	0001 0000	0100 0011
11	0001 0001	0100 0100
12	0001 0010	0100 0101

6. **Gray code:** It is another important non-weighted code. In gray code each number differs from its immediately preceeding number by one bit only.

7. Steps to be used to convert binary to gray code:

(*a*) Begin with the most significant bit (MSB). MSB of gray code is same as MSB of given binary number.

(*b*) The second MSB of gray code, is obtained by adding the MSB and the second MSB of binary number and ignoring carry if any.

(c) The third most significant bit (MSB) can be obtained by adding the third and second MSB of binary number and ignoring the carry if any. The process continues till we obtain the least significant bit (LSB) *e.g.,*

$$1 \quad 1 \quad 0 \quad 0 \quad 1 \quad 1 \rightarrow \text{binary}$$
$$\downarrow \quad \downarrow \quad \downarrow \quad \downarrow \quad \downarrow \quad \downarrow$$
$$1 \quad 0 \quad 1 \quad 0 \quad 1 \quad 0 \rightarrow \text{gray}$$

8. Steps to be used to convert gray code to binary:

(a) Begin with MSB. The MSB of the binary number is the same as MSB of the gray code number.

(b) The second MSB is obtained by adding the MSB in the binary number to the second MSB in the gray code number.

(c) Similarly the third MSB in binary number is obtained by adding second MSB in binary number to the third MSB in gray code number. The process continues till we obtain the LSB, *e.g.,*

$$1 \quad 0 \quad 1 \quad 0 \quad 1 \quad 0 \rightarrow \text{gray}$$
$$1 \quad 1 \quad 0 \quad 0 \quad 1 \quad 1 \rightarrow \text{binary}$$

9. **Alphanumeric codes:** To get information into and out of a computer, we need to use some kind of alphanumeric code (one for letters, numbers and other symbols).

There are two types of alphanumeric code:

(i) ASCII (American Standard Code for Information Interchange) code

(ii) EBCDIC (Extended Binary Coded Decimal Interchange Code).

10. **The ASCII code:** This code allows manufacturers to standardize computer hardware such as keyboards, printers and video displays. The ASCII code is a 7-bit code whose format is $X_6X_5X_4X_3X_2X_1X_0$ where each X is a 0 or a 1.

Table below shows to find the ASCII code for the uppercase and lowercase letters of the alphabet and some of the most commonly used symbols.

$X_3X_2X_1X_0$	$X_6X_5X_4$					
	010	011	100	101	110	111
0000	SP	0	@	P		p
0001	!	1	A	Q	a	q
0010	"	2	B	R	b	r
0011	#	3	C	S	c	s
0100	$	4	D	T	d	t
0101	%	5	E	U	e	u
0110	&	6	F	V	f	v
0111	'	7	G	W	g	w
1000	(	8	H	X	h	x

(Contd...)

$X_3X_2X_1X_0$	$X_6X_5X_4$					
	010	011	100	101	110	111
1001	)	9	I	Y	i	y
1010	*	:	J	Z	j	z
1011	+	;	K		k	
1100	,	<	L		l	
1101	-	=	M		m	
1110	.	>	N		n	
1111	/	?	O		o	

From the above table we find that capital letter A has an $X_6X_5X_4$ of 100 and an $X_3X_2X_1X_0$ of 0001. The ASCII code for A is therefore 1000001. Similarly the letter a is coded as 110 0001. SP stands for space (blank). Hitting the space bar of an ASCII keyboard sends this into a microcomputer 010 0000 (space).

10. **EBCDIC as alphanumeric code:** This is relatively less used alphanumeric code. This is an eight-bit code primarily used in IBM make devices. Here, the binary codes of letters and numerals come as an extension of BCD code. The bit assignments of EBCDIC are different from the ASCII but the character symbols are the same.

11. **Error Detecting codes:** In digital systems a word consisting of group of bits is stored as a unit. When this word is transmitted from one memory location to another or to an arithmetic unit, it is quite possible that an error may occur. This may happen due to any reason such as noise, by a transient, by an intermittent failure etc. This requires some method of detecting the errors. For detecting such errors, we use a simple method of parity.

12. **Parity code:** An error detection code using one additional parity bit.

13. **Parity bit:** A bit which is deliberately added to the binary number to make it either odd or even parity.

14. **Odd parity:** This means the adding of the parity bit to the group of bits (binary number) produce an odd number of 1's.

15. **Even parity:** This means the adding f the parity bit to the group of bits (binary number) produce an even number of 1's.

16. The ASCII code is used for sending digital data. 1-bit error may occur while transmission. To catch these errors, a parity bit is usually transmitted along with the original bits. Then a parity checker at the receiving end can test for even or odd parity, whichever parity has been prearranged between the sender and the receiver. Since ASCII code uses 7 bits, the addition of a parity bit to the transmitted data produces an 8-bit number in this format.

$$X_7X_6X_5X_4X_3X_2X_1X_0$$
$$\uparrow$$
Parity bit

This is an ideal length because most digital equipment is set up to handle bytes of data.

17. **Error correcting code or hamming code:** The parity check can detect an error which has crept in during the transmission of digital data *i.e.* parity bit indicates only that an error exists. It does not tell which bit is incorrect nor it corrects the incorrect bit. To overcome this problem another code called Hamming code is used. This code detects an error and indicates which bit is in error. This incorrect bit then can be changed to its correct form. This code is used for correcting a single error while transmission.

18. **Word format for Hamming code:** Let us take an example of 4 bit (BCD) data transmission. Here we use three check bits and the word format is:

$7\ 6\ 5\ 4\ 3\ 2\ 1\ \leftarrow$ Bit number

$D_7 D_6 D_5 P_4 D_3 P_2 P_1\ \leftarrow$ Allocated for

In the above format D represents the data bits and P represents the parity bits or check bits. Here D_7 is the MSB i.e. 4 bit word is $D_7 D_6 D_5 D_3$. The parity bit P_1, P_2 and P_4 are assigned values by making the following three parity relations each involving four out of seven bits.

(*a*) The first group involves the 1, 3, 5 and 7 bits (*i.e.*, $P_1 D_3 D_5 D_7$)

(*b*) The second group involves the 2, 3, 6 and 7 bits (*i.e.*, $P_2 D_3 D_6 D_7$)

(*c*) The third group involves the 4, 5, 6 and 7 bits (*i.e.*, $P_4 D_5 D_6 D_7$).

19. **Detecting the position of error in the data transmitted:** We can use odd-even parity check. If we use even parity, the first relation requires a value of P_1 such that the number of 1's in P_1, D_3, D_5 and D_7 is zero or even. The second relation gives the value of P_2 such that the number of 1's in P_2, D_3, D_6 and D_7 is zero or even. Similarly the third relation selects the value of P_4 such that the number of 1's in $P_4 D_5 D_6 D_7$ is zero or even.

In the data received, the error can be detected in any of the seven bit positions. Let us consider three bits x, y and z. The parity check is carried for all the above three parity relations. If the first parity relation is satisfied, $x = 0$ otherwise $x = 1$.

If the second parity relation is satisfied, $y = 0$ otherwise $y = 1$.

If the third parity relation is satisfied $z = 0$ otherwise $z = 1$.

The position of error is given by the three binary bit x, y and z and expressed as zyx, x being the LSB and z being the MSB.

Hence knowing the position of error, the data transmitted can be corrected by changing to 1 from 0 or from 0 to 1 in that position.

Thus seven bit Hamming code can be used for detecting error and requires extra transmission lines and extra digital circuitry for parity bit generator, error detector and error correcting circuits. This code can be used for more than 4 bit with the addition of more parity bits at each 2^n bit and holds for any length. For *e.g.*, a 16 bit data requires six check bits (parity bits) at 1, 2, 4, 8, 16, 32 place value for any single bit error.

20. **Checksum code:** An error detection code generating sum of a block of data.

21. **CRC code:** Cyclic Redundancy Code is a polynomial key based error detection code.

OBJECTIVE TYPE QUESTIONS

1. The examples of non weighted codes are _____ and _____.
2. The excess 3 code of 150_{10} is _____.
3. ASCII stands for _____.
4. EBCDIC stands for _____.
5. If there are odd number of 1's then parity is _____.
6. What is ASCII code?
7. What symbol is represented by the ASCII code 1000000.
8. What ASCII code is used for the percent sign, %?
9. What is the full form of CRC?
10. What is the advantage of Hamming code?
11. 1101 is valid BCD number (True or False).

Answers

1. Excess-3 code and gray code.
2. 0100 1000 0011.
3. American Standard Code for Information Interchange.
4. Extended Binary Code Decimal Interchange Code.
5. Odd parity.
6. ASCII code stands for American Standard code for Information Interchange code. A code which is used to represent alphanumeric information.
7. @
8. 010 0101.
9. Cyclic Redundancy code.
10. Hamming code can detect as well as correct one bit error.
11. False.

SHORT ANSWER TYPE QUESTIONS

Q.1. Convert the following decimal numbers to their 8421 BCD equivalents.

(*i*) 897, (*ii*) 214.83

Ans. (*i*) 8   9   7 → Decimal

      ↓   ↓   ↓

    1000 1001 0111 → BCD

∴   BCD equivalent is 100010010111 **Ans.**

(*ii*)   2   1   4 · 8   3 → Decimal

    ↓   ↓   ↓   ↓   ↓

   0010 0001 0100 · 1000 0011

∴   BCD equivalent is 001000010100.10000011 **Ans.**

Q.2. Convert binary number 10110001.11 to 8421 BCD equivalents.

Ans. To convert binary number to BCD equivalent first w have to convert the binary number into decimal form

∴ $10110001.11_2 = 1 \times 2^7 + 0 \times 2^6 + 1 \times 2^5 + 1 \times 2^4 + 0 \times 2^3 + 0 \times 2^2$

           $+ 0 \times 2^1 + 1 \times 2^0 + 1 \times 2^{-1} + 1 \times 2^{-2}$

          $= 128 + 0 + 32 + 16 + 0 + 0 + 0 + 1 + 0.5 + 0.25$

          $= 177.75_{10}$

Now   1   7   7 · 7   5 → Decimal

      ↓   ↓   ↓   ↓   ↓

    0001 0111 01111 · 0111 0101 → BCD

∴   BCD equivalent is 0001 01110111.01110101 **Ans.**

Q.3. What is the 4221 BCD equivalent of 98_{10}?

Ans. Decimal   →   9   8

              4221 4221

4221 BCD →   1111   1110

Hence 4221 BCD equivalent of decimal 98 is 11111110 **Ans.**

Q.4. What is the 5421 BCD equivalent of 98_{10}?

Ans. Decimal   →   9   8

              5421 5421

5421 BCD →   1100   1011

Hence 5421 BCD equivalent of decimal 98 is 11001011 **Ans.**

Q.5. 16 bit register contains the following

0100 1001 0101 0111 Interpret the contents as (*a*) decimal number, (*b*) two ASCII characters.

Ans. (*a*) 0100   1001   0101   0111

   ↓   ↓   ↓   ↓

   4   9   5   7

   ∴   Decimal equivalent is 4957_{10}

   (*b*) According to the ASCII code table

      0100 1001 → I   (ASCII code)

      0101 0111 → W   (ASCII code)

Q.6. Express decimal 5280 in excess-3 code

Ans. To convert the decimal number to excess-3 code, we have to add 3 to each digit and then find the BCD equivalent of each digit

$$
\begin{array}{cccc}
5 & 2 & 8 & 0 \\
+3 & +3 & +3 & +3 \\
\hline
8 & 5 & 11 & 3 \\
\downarrow & \downarrow & \downarrow & \downarrow \\
1000 & 0101 & 1011 & 0011
\end{array}
$$

   ∴ 5280_{10} in excess-3 code is 1000010110110011   **Ans.**

Q.7. Convert the XS3 number 1001 0101 1100 to its decimal equivalent.

Ans. To get the decimal equivalent from XS3 code we have to subtract 3 from each BCD number given in the XS3 number and then find the decimal equivalent for each BCD number

XS3 →   1001   0101   1100

          -0011 -0011 -0011

          0 1 1 0   0 0 1 0   1 0 0 1

            ↓   ↓   ↓

Decimal →   6   2   9

Hence the decimal equivalent of the given XS3 number is 629_{10} **Ans.**

Q.8. Add 36 and 39 in excess-3 code.

Ans.   36   0110   1001   →   Excess-3 equivalent of 36

          +39   0110   1100   →   Excess-3 equivalent of 39

          ───────────────────

          75   1101   0101   →   First result

Now the group which produces carry is in 8421 code and the group which accepts carry is in excess-6 code. Therefore add 3 to the group which produces carry and subtract 3 from the group which accepts carry to get the result in excess-3 code.

   ∴   1 0 1 0 1   0101   →   First result

             -0011 $+0011$

             ─────────────

             1010   1000   →   Excess -3 equivalent of 75

which is the required result.

Q.9. Add 42 and 34 in excess-3 code.

Ans.

42	0111	0101	→	Excess-3 equivalent of 42
+34	+0110	0111	→	Excess-3 equivalent of 34
76	1101	1100	→	Excess-6 equivalent of 76

Hence to get the result in exxess-3 code, subtract 3 from each group

∴   10101   10110   →   Excess-6 equivalent of 76

−0011   −0011

1010   1001   →   Excess -3 equivalent of 76

which is the required result.

Q.10. Convert 11001010_2 into gray code.

Ans.

1 1 0 0 1 0 1 0 → binary
↓ ↓ ↓ ↓ ↓ ↓ ↓ ↓
1 0 1 0 1 1 1 1 → gray

Q.11. Convert a gray code 10011011 to binary

Ans.

1 0 0 1 1 0 1 1 → gray
1 1 1 0 1 1 0 1 → binary

Q.12. What is the gray code for decimal 8

Ans. To get the gray code, first we have to convert decimal 8 to binary which is equal to 1000. Now from binary we will convert to gray

1 0 0 0 → binary
↓ ↓ ↓ ↓
1 1 0 0 → gray   **Ans.**

Q.13. With an ASCII keyboard, each keystroke produces the ASCII equivalent of the designated character. Suppose that you type PRINT X. What is the output of an ASCII keyboard.

Ans. The sequence is as follows:

P(101 0000), R(101 0010), I (100 1001)

N(100 1110), T(101 0100), space(010 0000)

X(101 1000)

Q.14. A computer sends a message to another computer using an odd-parity bit. Here is the message in ASCII codes plus the parity bit:

1100 1000

0100 0101

0100 1100

0100 1100

0100 1111

What do these numbers means.

Ans. First we notice that each 8-bit number has odd parity, an indication that no 1 bit error occurred during transmission. Next by using the ASCII table we find

100 1000	→	H
100 0101	→	E
100 1100	→	L
100 1100	→	L
100 1111	→	O

Therefore we get the message of HELLO.

Q.15. The transmitting computer sends the word GOODBYE. Show how this word is stored in the receiving computer. Use a starting address of 2000 and include a parity bit.

Ans. First let us write the ASCII code for each of the letter in GOODBYE with a parity bit to have odd parity

G	→	1100 0111
O	→	0100 1111
O	→	0100 1111
D	→	1100 0100
B	→	1100 0010
Y	→	1101 1001
E	→	0100 0101

Here is how the word is stored in the memory of the receiving computer.

Address	Alphanumeric	Hexadecimal contents
2000	G	C7
2001	O	4F
2002	O	4F
2003	D	C4
2004	B	C2
2005	Y	D9
2006	E	45

Q.16. Construct the even parity seven bit Hamming code for a word 1011.

Ans.

7	6	5	4	3	2	1	← Bit number
1	0	1	P_4	1	P_2	P_1	← Hamming code

Now to find the values of P_1, P_2 and P_4 we will use the three parity relations. From the first parity relation for bits 1, 3, 5 and 7 to have even parity P_1 should be 1.

From second parity relation for bits 2, 3, 6 and 7 to have even parity P_2 should be 0.

From the third parity relation for bits 4, 5, 6 and 7 to have even parity P_4 should be 0.

Hence the final code is

D_7	D_6	D_5	P_4	D_3	P_2	P_1
1	0	1	0	1	0	1

Q.17. A seven bit Hamming code is received as 1000010. What was the code transmitted or what s the correct code?

Ans. The code received

D_7	D_6	D_5	P_4	D_3	P_2	P_1
1	0	0	0	0	1	0

Applying first parity relation (for P_1, D_3, D_5 and D_7) we have $x = 1$ (as even parity is not observed). Applying second parity relation (for P_2, D_3, D_6 and D_7), we have $y = 0$ (as even parity is observed). Applying third parity relation (for P_4, D_5, D_6 and D_7), we have $z = 1$ (as even parity is not observed). Thus we find $zyx = 101$.

Hence the fifth data bit is having error and it should be corrected as 1 in place of 0. The correct code is therefore 1010010.

Q.18. A seven bit Hamming code coming out of transmission line is 0010100. Is there any error? If yes, in which data bit and what was the 4 bit data actually transmitted?

Ans. The code received

D_7	D_6	D_5	P_4	D_3	P_2	P_1
0	0	1	0	1	0	0

Applying first parity relation for P_1, D_3, D_5 and D_7 we have $x = 0$ (as even parity is observed). Applying second parity relation for P_2, D_3, D_6 and D_7 we have $y = 1$ (as even parity is not observed). Applying third parity relation for P_4, D_5, D_6 and D_7 we have $z = 1$ (as even parity is not observed).

Thus we find $zyx = 110$. Hence error is at sixth bit position. The correct bit is 1 in place of 0 as transmitted. Hence the data transmitted (D_7, D_6, D_5, D_3) was 0111.

Q.19. Encode data bits 1001 into a seven bit even parity Hamming code.

Ans.

7	6	5	4	3	2	1	← Bit number
1	0	0	P_4	1	P_2	P_1	← Hamming code

Using first parity relation for bit 1, 3, 5, 7 for even parity, P_1 should be 0. Using second parity relation for bit 2, 3, 6, 7 for even parity, P_2 should be 0 and using third parity relation for bit 4, 5, 6, 7 for even parity, P_4 should be 1.

Hence the 7 bit Hamming code for 4 bit data 1001 is 1001100. **Ans.**

Q.20. Explain the parity checker.

Ans. Exclusive –OR gates are ideal for checking the parity of a binary number since they produce an output 1 when the input has an odd number of 1's. Therefore an even parity input to an exclusive –OR 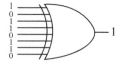 gate produces a low output while an odd parity input produces a high output. For *e.g.*, figure shows an 8 input Exclusive –OR gate. An eight bit number 10110110 drives the input. The exclusive –OR gate produces an output 1 because the input has odd parity (odd number of 1's). If the input changes to another value say 10110111, the output becomes 0 since the input is now even parity number (even number of 1's).

So, with the help of an Exclusive –OR gate we can check the parity of a binary number.

LONG ANSWER TYPE QUESTIONS

Q.1. Explain the odd-parity generator. What changes will you have to make in the circuit to make it even=parity generator?

Ans. In a computer, a binary number may represent an instruction that tells the computer to add, subtract and so on; or the binary number may represent data to be processed like a number, letter etc. In either case we sometimes see an extra bit added to the original binary number to produce a new binary number with even or odd parity. The circuit used to generate odd parity is shown below:

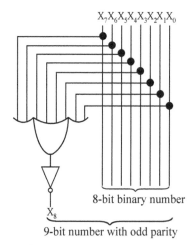

$X_7 X_6 X_5 X_4 X_3 X_2 X_1 X_0$

8-bit binary number

X_8

9-bit number with odd parity

Fig. shows 8 bit number $X_7 X_6 X_5 X_4 X_3 X_2 X_1 X_0$

Let this number be 0100 1101. Then the number has even parity. Hence the output of X-OR gate will be O. Again because of the inverter, $X_8 = 1$.

So the final 9 bit number is 1 0100 1101. Now we see that this has odd parity.

Now let us change the 8-bit input to 0110 1101. Now it has odd parity. Hence now the Exclusive OR gate will produce output 1. But the inverter produces 0 *i.e.*, $X_8 = 0$. Now the final 9-bit output is 0 0110 1101. Again the final output has odd parity.

Hence the above circuit is called odd parity generator because it always produces 9-bit output number with odd parity.

To get an even parity generator, the changes we have to make in the above circuit is that simply we have to delete the inverter. The output of the exclusive OR gate will be X_8 and now the 9-bit number will have even-parity.

Q.2. Explain how an error is detected in a data transmission system using parity bit.

Ans. In digital systems a word consisting of group of bits is stored as a unit and moves from one unit to another. It is quite possible that when this word is transmitted from one memory location to another or to an arithmetics unit, an error may occur. This may be due to any reason such as noise, by a transient, by an intermittent failure etc. This requires some method of detecting the errors. For *e.g.* it may be required to add 1100 0101 and 0100 1011. When the device transfers each of these words from the memory, it may transfer the first word as 1100 0100 (*i.e.* the first bit from the left of the first word is changed to 0 from 1). This will result in an error in the addition. For detecting such errors, we use a simple method of parity.

In this method, an addition bit known as parity bit is added with the numbers. The block diagram for a digital transmission system is shown below:

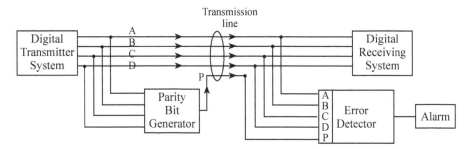

The transmitter is assumed to send a 4 bit word (A, B, C, D). These data bits are fed to a parity generator which sets the parity bit and this parity bit along with the four bits of data are sent through the transmission line and routed to the receiving system. Before the receiving system, the four bit word along with the parity bit is fed to a circuit called error-detection circuit. In case there has been some error during the transmission, this circuit activates the error alarm.

Odd as well as even parity are used in this method as per requirement of the digital system. The change in any bit will result in change in parity which can be detected on parity check. Of course in this method, we can detect the error only if there is change only in one bit.

Q.3. Explain how a double error is detected in a data transmission using double parity check.

Ans. If a word 0101 is transmitted as 1001 *i.e.*, first and second bit are interchanged then there are two errors—one in first bit and second in second bit. But we see that the number of 1's remain same and the parity check can not be successfully applied. To detect the double error we make use of dual parity or double parity check.

Double error normally occurs for the storage devices such as magnetic tapes. To detect the double errors, the dual parity check is done as follows: Figure below shows a portion of magnetic tape on which some information is stored:

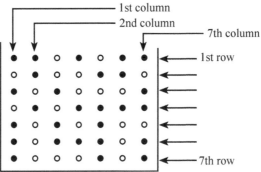

In the above figure by convention dark circle represent magnetised point as 1 bit and light dots represent unmagnetised point as 0 bit. Here in the tape 7 rows and 7 columns are shown. The seventh row and seventh column represents parity bit to make it odd parity when read horizontally or vertically. As shown in he example above when read horizontally

1st row   → 1101011 ⎫
2nd row   → 0100110 ⎬ odd parity
  ⋮
7th row   → 1000101 ⎭

When read vertically

1st column   → 1010111 ⎫
2nd column   → 1101000 ⎬ odd parity
  ⋮
7th column   → 1011011 ⎭

So, we find that either we read horizontally or vertically, all the rows and columns represent odd parity.

Now suppose while transmission or while storing double error occurs.

For *e.g.* in the above example if 1st bit if 1st row changes from 1 to 0, and 1st bit of 2nd row changes from 0 to 1, then when we read vertically.

1st column $\rightarrow$ 0110111 $\left.\right\}$ still odd parity
2nd column $\rightarrow$ 1101000

But now if we read horizontally

1st row $\rightarrow$ 0101011 $\left.\right\}$ even parity
2nd row $\rightarrow$ 1100110

Hence we find that when double error occurs, depending on the errors in columns or rows, reading horizontally or vertically (in one case) the parity changes and thus the double errors are detected by double-parity check.

7

Multiplexer, Demultiplexer, Encoder, Decoder

FACTS THAT MATTER

1. **Multiplexer:** Multiplexer means many into one A multiplexer is a circuit with many inputs but only one output. By applying control signals, we can steer any input to the output. Multiplexer is also called a data selector and control inputs are termed select inputs.

2. **Demultiplexer:** Demultiplexer means one into many. A demultiplexer is a logic circuit with one input and many outputs. By applying control signals, we can steer the input signal to one of the output lines.

3. **Data selector:** A synonym for multiplexer.

4. **Encoding:** The process of generating the binary codes from the decimal data for a digital system is called encoding.

5. **Encoder:** An encoder is a device that converts a decimal data to a coded output signal. An encoder has a number of inputs, only one of which is high at a time and an n-bit code is generated depending upon which of the input is excited. Encoder is similar to multiplexer.

6. **Decoding:** The process of converting the binary numbers or codes into decimal equivalent is called decoding.

7. **Decoder:** A decoder is similar to demultiplexer with the exception hat there is no data input. The only inputs are the control inputs. The control input bits produce one active output line. The decoder thus converts binary numbers or codes to its decimal equivalent. The most commonly used decoder is BCD-to-decimal decoder.

8. **Strobe:** An input that disables or enables a circuit.

9. **Active low:** The low state is the one that causes something to happen rather than the high state.

10. **Logic probe:** A troubleshooting device that indicates the state of a signal line.

OBJECTIVE TYPE QUESTIONS

1. Encoder does the reverse of _____.
2. Which of the following is used as a data selector?
   (*a*) Encoder, (*b*) Decoder, (*c*) Multiplexer, (*d*) Demultiplexer.
3. A circuit that transforms decimal number to binary code is _____.
4. A decoder has N inputs, this means there are _____ input codes.
5. A logic circuit which converts binary data to decimal data is _____.
6. Binary to octal decoder is also referred to as _____.
7. 1 of 10 decoder is _____ decoder.
8. There are 2^N input codes possible for an N bit input encoder (True or False).
9. What device will perform the conversion of a 4-bit BCD code to seven segment code required to drive a LED read out.
   (*a*) Decoder, (*b*) BCD to decimal decoder, (*c*) Multiplexer, (*d*) Demultiplexer.
10. Encoder are used in keyboards of calculator (True or False).

Answers

1. Decoder, **2.** Multiplexer, **3.** Encoder, **4.** 2^N, **5.** Decoder, **6.** 1 of 8 decoder
7. BCD-t0-decimal decoder, **8.** True, **9.** BCD-to-decimal decoder, **10.** True.

SHORT ANSWER TYPE QUESTIONS

Q.1. Draw he block diagram of a multiplexer. What is the basic function of a multiplexer?

Ans. A multiplexer is a circuit with many inputs but only one output. By applying control signals we can steer any input to the output. Thus it is also called a data selector. The block diagram of a multiplexer is shown below:

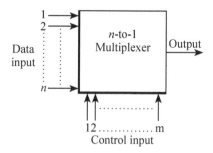

The block diagram shows that the circuit has n input signals, m control signals and 1 output signal. The control inputs are also termed select inputs. m control signals can select at the most 2^m input signals. Thus $n \le 2^m$. Now the basic function of a multiplexer is that with the help of the control inputs, we can steer the desired data input to the output. The number of bits of the control inputs depend on the number of input signals: For *e.g.* for a 2 : 1 MUX (multiplexer), control input m = 1 as $2 = 2^m$ or $m = 1$.

For 4 : 1 MUX, $n = 4$ ∴ $4 = 2^m$ or $2^2 = 2^m$ or $m = 2$

For 8 : 1 MUX, $n = 8$ ∴ $8 = 2^m$ or $2^3 = 2^m$ or $m = 3$ and so on.

Q.2. Describe a two-input multiplexer?

Ans. Since there are two inputs, hence $n = 2$

∴ $2 = 2^m$ or $2^1 = 2^m$ or $m = 1$ where m is the control input.

So, for a 2 input multiplexer we need only 1 control input.

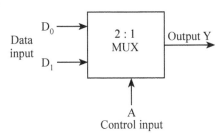

Block diagram of 2 : 1 MUX

The circuit diagram of 2-to-1 multiplexer will be as below (Fig. a):

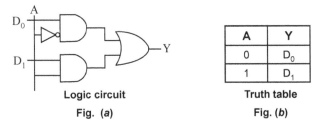

A	Y
0	D_0
1	D_1

Logic circuit   **Truth table**

Fig. (a)   **Fig. (b)**

Depending on the control input A, one of the two inputs D_0 or D_1 is steered to output Y. Let us write the logic equation of the circuit of Fig. (a)

$$Y = A'D_0 + AD_1$$

Now there will be two possible conditions:

Case 1:   If A = 0,

$$Y = 1.D_0 + 0.D_1 = D_0$$

Case 2:   If A = 1,

$$Y = 0.D_0 + 1.D_1 = D_1$$

In other words if control input A = 0, the first AND gate to which D_0 is connected remains active and equal to D_0 while the other AND gate is inactive. Thus multiplexer output Y is same as D_0. If $D_0 = 0$, Y = 0 and if $D_0 = 1$, Y = 1.

Similarly for control input A = 1, the second AND gate will be active the other AND gate will be inactive. Thus output $Y = D_1$.

Hence the truth table will be as shown in Fig. (b) above for a 2 : 1 multiplexer.

Q.3. Explain the working of a 4-to-1 multiplexer?

Ans. 4-to-1 multiplexer means there are 4 inputs and 1 output.

Since there are for inputs, hence $n = 4$

∴ $4 = 2^m$ or $2^2 = 2^m$ or $m = 2$ where m is the control inputs.

So, for a 4 input multiplexer we need 2 control signals.

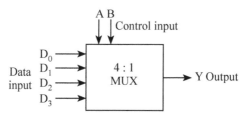

Black diagram o 4 : 1 MUX

The circuit diagram of 4-to-1 multiplexer is shown in Fig. (a) below:

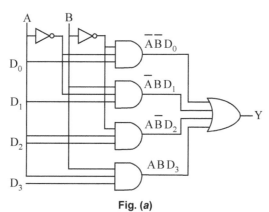

Fig. (a)

In the above circuit the two required control inputs are A and B. The four inputs are D_0 to D_3. Now depending on the values of control inputs one of the four inputs will be steered to the output Y.

Let us write the logic equation of the circuit.

$$Y = \overline{A}\,\overline{B}D_0 + \overline{A}BD_1 + A\overline{B}D_2 + ABD_3.$$

Since there are two control inputs, hence there will be 4 possible cases

Case 1: $A = 0, B = 0$ then $Y = \overline{0}\ \overline{0}\ D_0 + \overline{0}.0D_1 + 0.\overline{0}D_2 + 0.0D_3$
   or $Y = 1.1D_0 + 0 + 0 + 0 = D_0$
Case 2: $A = 0, B = 1$ then $Y = \overline{0.1}\ D_0 + \overline{0}.1D_1 + 0.\overline{1}D_2 + 0.1D_3$
   or $Y = 0 + 1.1D_1 + 0 + 0 = D_1$
Case 3: $A = 1, B = 0$ then $Y = \overline{1.0}\ D_0 + \overline{1}.0D_1 + 1.\overline{0}D_2 + 1.0D_3$
   or $Y = 0.1D_0 + 0.0D_1 + 1.1D_2 + 0$
   $= 0 + 0 + D_2 + 0 = D_2$
Case 4: $A = 1, B = 1$ then $Y = \overline{1.1}\ D_0 + \overline{1}.1D_1 + 1.\overline{1}D_2 + 1.1D_3$
   or $Y = 0.0D_0 + 0.1D_1 + 1.0D_2 + D_3$
   $= 0 + 0 + 0 + D_3 = D_3$
Hence the truth table of 4 : 1 MUX is as follows

A	B	C
0	0	D_0
0	1	D_1
1	0	D_2
1	1	D_3

Truth Table

In other words, for $AB = 00$, the first AND gate to which D_0 is connected remains active and equal to D_0 while the other AND gates are inactive. Thus multiplexer output is same as D_0. If $D_0 = 0$, $Y = 0$, if $D_0 = 1$, $Y = 1$.

Similarly, if $AB = 01$, only second AND gate will be active to which D_1 is connected and all he other AND gates remain inactive. Thus now output $Y = D_1$. Following the same procedure the truth table for 4 : 1 multiplexer is completed as shown above.

Q.4. Show how 4-to-1 multiplexer can be obtained using only 2-to-1 multiplexer.

Ans. Logic equation for 2-to-1 multiplexer is

$$Y = A'D_0 + AD_1 \qquad \qquad ...(i)$$

where A is the control input and D_0, D_1 are data inputs.
Logic equation for 4-to-1 multiplexer is

$$Y = A'B'D_0 + A'BD_1 + AB'D_2 + ABD_3 \qquad \qquad ...(ii)$$

where A, B are control inputs and D_0 to D_3 are data inputs.
Now equation (ii) can be rewritten as

$$Y = A'(B'D_0 + BD_1) + A(B'D_2 + BD_3) \qquad \qquad ...(iii)$$

Now, if we compare equation (iii) with equation (i) we find that we need two 2-to-1 multiplexer to realise two bracketed terms where B serves as the control input. The output of thee two multiplexers can be sent to a third 2-to-1 multiplexer as data inputs where A serves as control input and thus we get 4-to-1 multiplexer using only 2-to-1 multiplexer. The block diagram is shown below

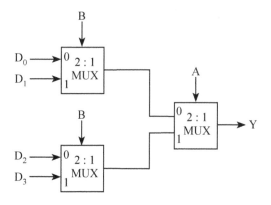

LONG ANSWER TYPE QUESTIONS

Q.1. (*a*) Realise $Y = A'B + B'C' + ABC$ using 8-to-1 multiplexer.

(*b*) Can it be realised with a 4-to-1 multiplexer

Ans. (a) First we express Y as a function of minterms of three variables. Thus

$Y = A'B + B'C' + ABC$

$= A'B(C + C') + B'C'(A + A') + ABC$ $(\because \quad X + X' = 1)$

$= A'BC + A'BC' + AB'C' + A'B'C' + ABC$

$= A'B'C' + A'BC' + A'BC + AB'C' + ABC$ ...(i)

Now, logic equation for 8-to-1 multiplexer is

$Y = A'B'C' \, D_0 + A'BC \, D_1 + A'BC' \, D_2 + A'BC \, D_3 + AB'C' \, D_4$

$\quad + AB'C \, D_5 + ABC' \, D_6 + ABCD_7$ ...(ii)

Comparing equation (i) with equation (ii) we find by substituting $D_0 = D_2 = D_3 = D_4 = D_7 = 1$ and $D_1 = D_5 = D_6 = 0$, we get given logic relation. Thus the given equation can be realised using 8-to-1 multiplexer.

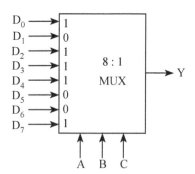

(*b*) For 4-to-1 multiplexer, we need two control inputs. Let A and B be used as control inputs and C fed as input. The 4-to-1 MUX generates

4 minterms for different combination of AB. Let us now rewrite the given logic equation in such as way that all these terms are present in the equation. Thus

$Y = A'B + B'C' + ABC$

$= A'B + B'C' (A' + A) + ABC \quad (\because X + X' = 1)$

$= A'B + A'B'C' + AB'C' + ABC$

$= A'B'.C' + A'B.1 + AB'.C' + AB.C \qquad \qquad ...(i)$

Now logic equation for 4-to-1 multiplexer is

$$Y = A'B.D_0 + A'B.D_1 + AB'.D_2 + AB'.D_3 \qquad ...(ii)$$

Comparing equation (i) with equation (ii) we find that $D_0 = C'$, $D_1 = 1$, $D_2 = C'$ and $D_3 = C$ generate the given logic function. Thus the given equation can be realised using 4-to1 multiplexer

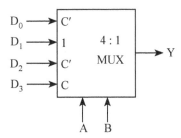

Q.2. Explain the working of 8-to-1 multiplexer. Also give the pin diagram, logic diagram and truth table of 8 : 1 multiplexer IC74151.

Ans. 8-to-1 multiplexer means there are 8 inputs and 1 output. Since there are 8 inputs, hence $n = 8$. Therefore $8 = 2^m$ or $2^3 = 2^m$ or $m = 3$ where m is the control inputs., Since we need 3 control inputs, so 8 : 1 multiplexer is sometimes called 8 : 3 multiplexer also.

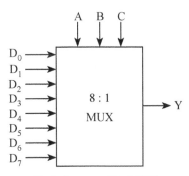

Block diagram of 8 : 1 MUX

The circuit diagram of 8-to-1 multiplexer is shown below:

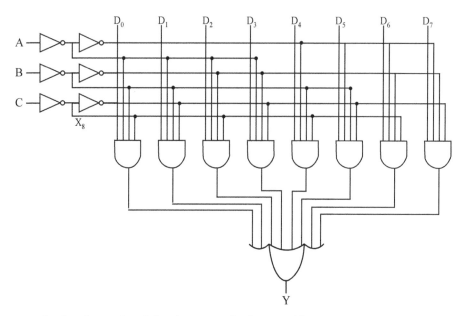

In the above circuit he three required control inputs are A, B and C. The eight inputs are D_0 to D_7. Now depending on the values of control inputs, one of the eight inputs will be steered to the output Y.

Let us write the logic equation of the circuit

$$Y = A'B'C'D_0 + A'B'CD_1 + A'BC'D_2 + A'BCD_3 + AB'C'D_4 + AB'CD_5 + ABC'D_6 + ABCD_7$$

Since there are three control inputs, hence there will be eight possible cases.

Case 1: A = 0, B = 0, C = 0, then

$$Y = 0'0'0'D_0 + 0'0'0D_1 + 0'00'D_2 + 0'.0.0D_3 + 0.0'0'D_4 + 0.0'0D_5 + 0.0.0'D_6 + 0.0.0D_7$$

$$= 1.1.1.D_0 + 0 + 0 + 0 + 0 + 0 + 0 + 0$$

$$= D_0 \therefore \text{ if } D_0 = 0, Y = 0. \text{ If } D_0 = 1, Y = 1.$$

Similarly for the other seven cases i.e. for A, B, C as 001, 010, 011, 100, 101, 110 and 111 we will have $Y = D_1, D_2, D_3, D_4, D_5, D_6$ and D_7 correspondingly.

In other words for ABC = 000, the first AND gate to which D_0 is connected remains active and equal to D_0 while the other AND gates are inactive. Thus multiplexer output is same as D_0. If $D_0 = 0$, Y = 0, if $D_0 = 1$, Y = 1 irrespective of other input values. Similarly the output of multiplexer can be explained for a particular control inputs. So we find that at a time only one data input is steered to the output Y depending on the control inputs.

Figure below show the pin diagram and logic diagram of IC74151 which is 8 : 1 multiplexer

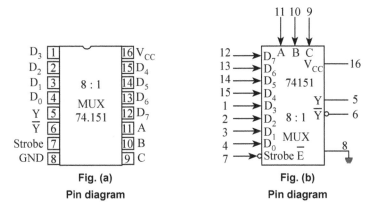

Fig. (a)

Pin diagram

Fig. (b)

Pin diagram

IC74151 is a 8 : 1 MUX whose pin diagram is shown in Fig. (a). It is 16 pin IC. The logic diagram is shown in Fig. (b). Here we find that pin no. 16 is for V_{CC} and pin 8 is grounded. The inputs D_0 to D_7 are connected to the pins as shown above. The control inputs are A, B, C connected to pin 11, 10 and 9 correspondingly. The output Y is connected to pin 5 and we also get complemented output $\overline{Y}$ which is connected to pin 6. Now pin 7 is for active low strobe signal. If strobe is low, then only the multiplexer is enabled and the MUX functions and we get the desired output. If strobe is high, then we get 0 output irrespective of the values of control inputs or data inputs. The truth table of 8 : 1 MUX (IC74151) is shown below:

Strobe	Control inputs			Data inputs								Output	
$\overline{E}$	A	B	C	D_0	D_1	D_2	D_3	D_4	D_5	D_6	D_7	Y	$\overline{Y}$
1	X	X	X	X	X	X	X	X	X	X	X	0	1
0	0	0	0	0	X	X	X	X	X	X	X	0	1
0	0	0	0	1	X	X	X	X	X	X	X	1	0
0	0	0	1	X	0	X	X	X	X	X	X	0	1
0	0	0	1	X	1	X	X	X	X	X	X	1	0
0	0	1	0	X	X	0	X	X	X	X	X	0	1
0	0	1	0	X	X	1	X	X	X	X	X	1	0
0	0	1	1	X	X	X	0	X	X	X	X	0	1
0	0	1	1	X	X	X	1	X	X	X	X	1	0
0	1	0	0	X	X	X	X	0	X	X	X	0	1
0	1	0	0	X	X	X	X	1	X	X	X	1	0
0	1	0	1	X	X	X	X	X	0	X	X	0	1
0	1	0	1	X	X	X	X	X	1	X	X	1	0
0	1	1	0	X	X	X	X	X	X	0	X	0	1
0	1	1	0	X	X	X	X	X	X	1	X	1	0
0	1	1	1	X	X	X	X	X	X	X	0	0	1
0	1	1	1	X	X	X	X	X	X	X	1	1	0

In the above truth table X → don't care.

Q.3. (*a*) Implement the expression $f(X, Y, Z) = \Sigma m(1, 3, 5, 6, 7)$ using a multiplexer.

(*b*) What are the advantages of multiplexer?

Ans. (*a*) The given expression is

$$f(X, Y, Z) = \Sigma m(1, 3, 5, 6, 7)$$

Here we see that number of variables are 3. Therefore a multiplexer with three control inputs is required i.e. we need a 8 : 1 multiplexer (IC74151). Circuit of IC74151 (8 : 1 MUX) connected to implement the given expression is show in Figure below.

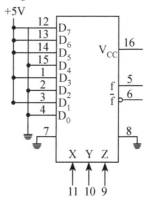

Based on the given expression D_1, D_3, D_5, D_6 and D_7 lines are connected to a HIGH or logic 1 and D_0, D_2 and D_4 are connected to LOW or logic 0. When we sequence the data select inputs from 000 to 111, the required Boolean function will be generated at the output in a serial form as indicated in the truth table for this circuit as shown below:

Control inputs	Data	Output
X Y Z	Line	Value
0 0 0	D_0	0
0 0 1	D_1	1
0 1 0	D_2	0
0 1 1	D_3	1
1 0 0	D_4	0
1 0 1	D_5	1
1 1 0	D_6	1
1 1 1	D_7	1

(*b*) **Advantages of multiplexer:** Some of the advantages of mutiplexer are as follows

(*i*) A multiplexer logic circuit can be used to generate random logic functions. This can also be done using logic gates, but it is more advantageous to use a mutiplexer for this purpose.

(*ii*) For implementing any logic function, no simplification of Boolean expression is required.

(*iii*) It minimizes the IC package count

(*iv*) Logic design is much simplified using multiplexer.

(*v*) Multiplexer is also used for parallel to serial conversion. Multiplexers take data is parallel from several sources and send the data out serially on one (single) line in a time sequence.

(*vi*) Multiplexer is mainly used as data selector (data routing).

Q.4. Explain the working of 16 : 1 multiplexer.

Ans. 16 : 1 multiplexer means there are 16 inputs and 1 output. Since there are 16 inputs, hence $n = 16$. Therefore $16 = 2^m$ or $2^4 = 2^m$ or $m = 4$ where m is the control inputs. The block diagram of 16 : 1 MUX is shown below:

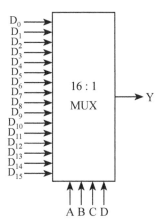

The circuit diagram of 16-to1 MUX is shown below:

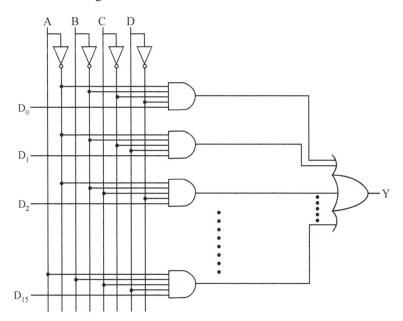

In the above circuit the four required control inputs are A, B, C and D. The 16 inputs are D_0 to D_{15}. Now depending on the values of control inputs, one of the sixteen inputs will be steered to the output Y.

Let us write the logic equation of the circuit

$Y = A'B'C'D'.D_0 + A'B'C'D.D_1 + A'B'CD'.D_2 + A'B'CD.D_3 + A'BC'D'.D_4 +$
$\quad A'BC'D.D_5 + A'BCD'.D_6 + A'BCD.D_7 + AB'C'D'.D_8 + AB'C'D.D_9 +$
$\quad AB'CD'.D_{10} + AB'CD.D_{11} + ABC'D'.D_{12} + ABC'D.D_{13} + ABCD'.D_{14}$
$\quad + ABCD.D_{15}$

Since there are four control inputs, hence there will be sixteen possible cases.

Case 1: A = 0. B = 0, C = 0, D = 0 then

$Y = 0'0'0'0'.D_0 + 0 + 0 + 0 + 0 + 0 + 0 + 0 + 0 + 0 + 0 + 0 + 0 + 0 + 0$

$\quad = 1.1.1.1.D_0 = D_0.$

Similarly for the other fifteen cases, we will have $Y = D_1, D_2, ... D_{15}$ correspondingly. In other words for ABCD = 0000, the first AND gate to which D_0 is connected remains active and output equals to D_0 while the other AND gates are inactive. Thus multiplexes output is same as D_0. If $D_0 = 0$, $Y = 0$, if $D_0 = 1$, $Y = 1$ irrespective of other input values. Similarly the output of multiplexer can be explained for all the other possible control inputs, the output will be according to the corresponding data input attached to that particular AND gate.

So we find that at a time only one data input is steered to the output Y depending on the control inputs.

Q.5. Draw and explain the pin diagram of IC74150. Draw the logic diagram from its truth table.

Ans. If we try to visualize the 16-input OR gate of the 16-to-1 multiplexer changed to a NOR gate, we get the complement of the selected data bit rather than the data bit-itself. For e.g. when control inputs ABCD = 0110, the output is

$$Y = \overline{D_6}$$

This is the Boolean equation for a typical transistor-transistor logic (TTL) multiplexer because it has an inverter on the output that produces the complement of the selected data bit.

The 74150 is a 16-to-1 TTL multiplexer IC with the pin diagram shown in Fig. (a) below. Pins 1 to 8 and 16 to 23 are for the input data bits D_0 to D_{15}. Pins 11, 13, 14 and 15 are for the control bits ABCD. Pin 10 is the output and it equals the complement of the selected data bit. Pin 9 is for the strobe, an input signal that disables or enables the multiplexer. A low strobe enables the MUX so that output Y equals the complement of the input data bit $Y = \overline{D_n}$.

where n is the decimal equivalent of ABCD. On the other hand, a high strobe disables the multiplexer and forces the output into the high state and in this case the value of control inputs ABCD doesn't matter. The truth table is shown in Fig. (b) below. 74150 is a 24 pin IC.

D_7 [1]	74150	[24] V_{CC}
D_6 [2]		[23] D_8
D_5 [3]		[22] D_9
D_4 [4]		[21] D_{10}
D_3 [5]		[20] D_{11}
D_2 [6]		[19] D_{12}
D_1 [7]		[18] D_{13}
D_0 [8]		[17] D_{14}
Strobe [9]		[16] D_{15}
Y [10]		[15] D
A [11]		[14] C
GND [12]		[13] B

Fig. (a)
Pin diagram

Strobe	A	B	C	D	Y
L	L	L	L	L	$\overline{D}_0$
L	L	L	L	H	$\overline{D}_1$
L	L	L	H	L	$\overline{D}_2$
L	L	L	H	H	$\overline{D}_3$
L	L	H	L	L	$\overline{D}_4$
L	L	H	L	H	$\overline{D}_5$
L	L	H	H	L	$\overline{D}_6$
L	L	H	H	H	$\overline{D}_7$
L	H	L	L	L	$\overline{D}_8$
L	H	L	L	H	$\overline{D}_9$
L	H	L	H	L	$\overline{D}_{10}$
L	H	L	H	H	$\overline{D}_{11}$
L	H	H	L	L	$\overline{D}_{12}$
L	H	H	L	H	$\overline{D}_{13}$
L	H	H	H	L	$\overline{D}_4$
L	H	H	H	H	$\overline{D}_{15}$
H	X	X	X	X	H

Fig. (b)
Truth table of 74150

To draw the logic diagram of IC 74150 we need to analyse the truth table in Fig. (b). We find that strobe is an active-low signal i.e. it causes something to happen when it is low rather than when it is high. To indicate this bubble is used in the logic diagram. Similarly we see that the output Y is complemented. So the output line is also represented by a bubble. The logic diagram of IC 74150 is now shown below:

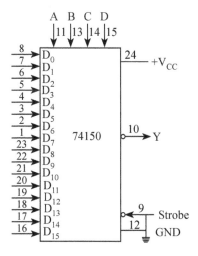

Q.6. Making use of appropriate multiplexer implement the following logic function.

$$F = AB\overline{C}\,\overline{D} + A\overline{B}\,\overline{C}D + \overline{A}\,\overline{B}C + \overline{A}BC$$

Ans. $F = AB\overline{C}\,\overline{D} + A\overline{B}\,\overline{C}D + \overline{A}\,\overline{B}C + \overline{A}BC$

$$= AB\overline{C}\,\overline{D} + A\overline{B}\,\overline{C}D + \overline{A}\,\overline{B}C(D + \overline{D}) + \overline{A}BC(D + \overline{D}) \quad \because\ X + \overline{X} = 1$$

$$= AB\overline{C}\,\overline{D} + A\overline{B}\,\overline{C}D + \overline{A}\,\overline{B}CD + \overline{A}\,\overline{B}C\overline{D} + \overline{A}BCD + \overline{A}BC\overline{D}$$

Now first we have to write the Truth Table for the given expression in which output will be 1 for the terms present in the expression and 0 for the other possible terms not present in the given expression. Accordingly the truth table is shown below in Fig. (a). Since there are four variables hence there will be 16 possible combinations.

Input				Decimal equivalent	Output
A	B	C	D		
0	0	0	0	0	0
0	0	0	1	1	0
0	0	1	0	2	1
0	0	1	1	3	1
0	1	0	0	4	0
0	1	0	1	5	0
0	1	1	0	6	1
0	1	1	1	7	1
1	0	0	0	8	0
1	0	0	1	9	1
1	0	1	0	10	0
1	0	1	1	11	0
1	1	0	0	12	1
1	1	0	1	13	0
1	1	1	0	14	0
1	1	1	1	15	0

If A, B, C and D variables are taken as the control inputs of a 16-to-1 multiplexer (with four variables we have 16 possible combinations) and the appropriate digital levels are placed at the multiplexer data inputs, we can implement the function F.

We make use of IC74150 (16-to-1 MUX). At each data input that satisfies any term in the Boolean expression given, a 1 must be placed. The connection according to the truth table to implement the function F is shown in the figure below. To ensure the working of this circuit, let us consider some entries from truth table, *e.g.* for ABCD = 0010, the multiplexer will select D_2 which is 1. This gets complemented twice before reading F, so F = 1 which is correct. If ABCD = 0011, the MUX will select D_3 which is 1. This gets complemented twice before reaching F; so F = 1 which is correct. Apart from the combinations of ABCD given in the expression F will be zero for all other possible combinations of ABCD, which is correct.

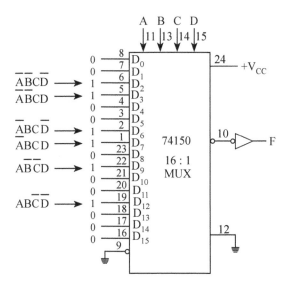

Q.7. What is a demultiplexer? Draw the block diagram and explain the basic function of a demultiplexer.

Ans. Demultiplexer means one into many. A demultiplexer is a logic circuit with one input and many outputs. By applying control signals, we can steer the input data to one of the output lines. The block diagram of demultiplexer is shown below:

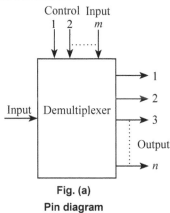

Fig. (a)
Pin diagram

The circuit has 1 inputs signal, m control or select signals and n output signal where $n \le 2^m$. Now the basic function of a demultiplexer is that with the help of the control signals or inputs, we can steer the input data to the desired output line. The number of bits of the control signal depend on the number of output lines, for e.g. a 1-to-2 demultiplexes will have control input $m = 1$ as $2 = 2^m$ or $m = 1$.

For 1-to-4 demultiplexer $n = 4$ ∴ $4 = 2^m$ or $2^2 = 2^m$ or $m = 2$

For 1-to-8 demultiplexer $n = 8$ ∴ $8 = 2^m$ or $2^3 = 2^m$ or $m = 3$ and so on.

To explain the function of demultiplexer let us take an example of 1-to-2 demultiplexer. Figure below shows the circuit diagram of 1-to-2 demultiplexer.

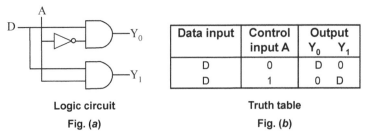

Data input	Control input A	Output Y₀ Y₁	
		Y$_0$	Y$_1$
D	0	D	0
D	1	0	D

Logic circuit

Fig. (a)

Truth table

Fig. (b)

Since there are two output lines, hence $n = 2$

∴ $2 = 2^m$ or $2^1 = 2^m$ or $m = 1$. Here A is the control input. Now depending on control input A, the data input D is steered to the output line Y or Y1 of Fig. (a).

If $A = 0$, w find that the upper AND gate will be enabled and lower AND gate will be disabled. Hence D input will steer to Y_0 or $Y_0 = D$. If $D = 0$, $Y_0 = 0$. If $D = 1$, $Y_0 = 1$. If $A = 1$, the lower AND gate will be enabled and upper AND gate will be disabled. Hence now D input will steer to Y_1 or $Y_1 = D$. If $D = 0$, $Y_1 = 0$, if $D = 1$, $Y_1 = 1$.

The truth table of 1-to-2 demultiplexer is shown in Fig. (b).

Q.8. With a neat diagram explain 1 : 8 demultiplexer and realise it using gates.

Ans. A 1 : 8 demultiplexer has one input signal and eight output signals. The number of control inputs required is $8 = 2^m$ or $2^3 = 2^m$ or $m = 3$ (where m is the required control inputs). The block diagram is shown below:

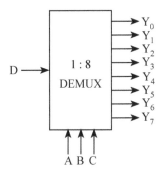

As shown in the block diagram A, B and C are the three required control inputs D is the data input and Y_0 to Y_7 are the eight output lines. The data bit (input) is steered to the output line whose subscript is the decimal equivalent of ABC.

Since there are three control inputs, there will be eight possible cases

Case 1: $A = 0$, $B = 0$, $C = 0$, then $Y_0 = D$

Case 2: $A = 0$, $B = 0$, $C = 1$, then $Y_1 = D$

Case 3: $A = 0$, $B = 1$, $C = 0$, then $Y_2 = D$

Case 4: $A = 0$, $B = 1$, $C = 1$, then $Y_3 = D$

Case 5: $A = 1$, $B = 0$, $C = 0$, then $Y_4 = D$

Case 6: $A = 1$, $B = 0$, $C = 1$, then $Y_5 = D$

Case 7: $A = 1$, $B = 1$, $C = 0$, then $Y_6 = D$

Case 8: $A = 1$, $B = 1$, $C = 1$, then $Y_7 = D$

The above function of the 1 : 8 Demux can be realised by the logic circuit using gates as follows:

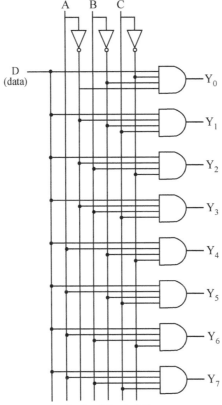

Logic circuit for 1: 8 Demux

In the above circuit the three required control inputs are A, B and C. The eight output lines are Y_0 to Y_7. D is the data input. Now depending on the values of the control inputs, the data D will be steered to one of the output lines. Let, $A = 0$, $B = 0$, $C = 0$, then according to the circuit we find only the first AND gate is enabled while all other AND gates will be disabled. Thus in this case, input D will steer to Y_0, i.e. $Y_0 = D$. If D

$= 0$, $Y_0 = 0$, if $D = 1$, Y_0 will be 1 while all other outputs will be zero irrespective of the value of D.

Similarly if $A = 0$, $B = 0$, $C = 1$, then the second AND gate whose output is Y_1, will be enabled while all other AND gates will be disabled. Thus now D will be steered to Y_1 or $Y_1 = D$. If $D = 0$, $Y_1 = 0$. If $D = 1$, $Y_1 = 1$ while all other outputs will remain 0 irrespective of the value of D.

Similarly we can explain for all the other six AND gates. Always we will find that depending on the values of control inputs A, B and C, only one AND gate will be enabled and input D will steer to that output line.

Q.9. (*a*) Explain the working of 1 : 16 demultiplexer.

   (*b*) Draw and explain the pin diagram of IC74154 and draw its logic diagram from the truth table.

Ans. (*a*) A 1 : 16 demultiplexer has one input signal and sixteen output signals. The number of control inputs required here is $16 = 2^m$ or $2^4 = 2^m$ or $m = 4$ (where *m* is the required control inputs). The block diagram is shown below:

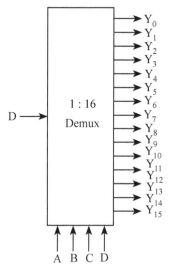

As shown in the block diagram A, B, C and D are the required control inputs. D is the data input and Y_0 to Y_{15} are the 16 output lines. The data bit is steered to the output line whose subscript is the decimal equivalent of ABCD.

The circuit for 1: 16 demultiplexer using gates is shown below:

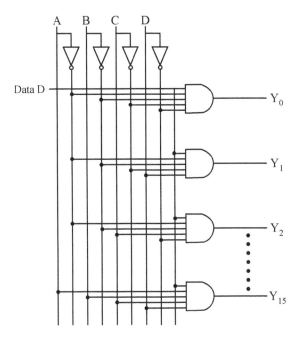

In the above circuit, when ABCD = 0000, the upper AND gate is enabled while all other AND gates are disabled. Therefore data D is transmitted only to the Y_0 output giving $Y_0 = D$. If D is low, Y_0 is low. If D is high Y_0 is high. As we can see, in this case the value of Y_0 depends on the value of D. All other outputs are in the low state. If control input is changed to ABCD = 1111, all the gates are disabled except the bottom AND gate. Then D is transmitted only to the Y_{15} output and $Y_{15} = D$. Similarly we can explain for any other value of ABCD and find that at a time only one AND gate is enabled and the data bit D is transmitted to the output line whose subscript is the decimal equivalent of ABCD.

Ans. (b) IC74154 is a 1-to-16 demultiplexer. The pin diagram is shown below. It is a 24 pin IC. Pin 18 is for the input DATA D, pins 20 to 23 are for the control input ABCD. Pin 1 to 11 and 13 to 17 are for the output bits Y_0 to Y_{15}. Pin 19 is for the STROBE, an active low input. Finally pin 24 is for V_{CC} and pin 12 for ground.

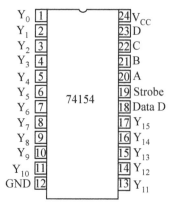

Pin Diagram of 74154

Truth Table of 74154

Strobe	Data	A	B	C	D	Y₀	Y₁	Y₂	Y₃	Y₄	Y₅	Y₆	Y₇	Y₈	Y₉	Y₁₀	Y₁₁	Y₁₂	Y₁₃	Y₁₄	Y₁₅
L	L	L	L	L	L	L	H	H	H	H	H	H	H	H	H	H	H	H	H	H	H
L	L	L	L	L	H	H	L	H	H	H	H	H	H	H	H	H	H	H	H	H	H
L	L	L	L	H	L	H	H	L	H	H	H	H	H	H	H	H	H	H	H	H	H
L	L	L	L	H	H	H	H	H	L	H	H	H	H	H	H	H	H	H	H	H	H
L	L	L	H	L	L	H	H	H	H	L	H	H	H	H	H	H	H	H	H	H	H
L	L	L	H	L	H	H	H	H	H	H	L	H	H	H	H	H	H	H	H	H	H
L	L	L	H	H	L	H	H	H	H	H	H	L	H	H	H	H	H	H	H	H	H
L	L	L	H	H	H	H	H	H	H	H	H	H	L	H	H	H	H	H	H	H	H
L	L	H	L	L	L	H	H	H	H	H	H	H	H	L	H	H	H	H	H	H	H
L	L	H	L	L	H	H	H	H	H	H	H	H	H	H	L	H	H	H	H	H	H
L	L	H	L	H	L	H	H	H	H	H	H	H	H	H	H	L	H	H	H	H	H
L	L	H	L	H	H	H	H	H	H	H	H	H	H	H	H	H	L	H	H	H	H
L	L	H	H	L	L	H	H	H	H	H	H	H	H	H	H	H	H	L	H	H	H
L	L	H	H	L	H	H	H	H	H	H	H	H	H	H	H	H	H	H	L	H	H
L	L	H	H	H	L	H	H	H	H	H	H	H	H	H	H	H	H	H	H	L	H
L	L	H	H	H	H	H	H	H	H	H	H	H	H	H	H	H	H	H	H	H	L
L	H	X	X	X	X	H	H	H	H	H	H	H	H	H	H	H	H	H	H	H	H
H	L	X	X	X	X	H	H	H	H	H	H	H	H	H	H	H	H	H	H	H	H
H	H	X	X	X	X	H	H	H	H	H	H	H	H	H	H	H	H	H	H	H	H

From the truth table of 74154, we notice that the strobe input must be low to activate the Demux. As long as the STROBE is low, the control input ABCD determines which output line is low when the DATA input is low. When DATA input is high, all output lines are high and when STROBE is high, all output lines are high.

Figure below shows the schematic diagram or logic diagram of IC74154.

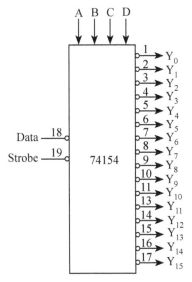

Logic diagram of 74154

There is one DATA bit (pin 18) under the control of nibble ABCD. The DATA bit is automatically steered to the output line whose subscript is the decimal equivalent of ABCD. Again the bubble on the STROBE pin indicates an active-low input. We find from the logic diagram that DATA is inverted at the input (shown by the bubble on pin 18) and again on any output (the bubble on each output pin). With double inversion, DATA passes through the 74154 unchanged.

Q.10. What is a decoder? Explain the working of BCD-to-decimal decoder. Draw the pin diagram, logic diagram and write the truth table for IC7445 which is BCD-to-decimal decoder.

Ans. A decoder is similar to demultiplexer with the exception that there is no data input. The only inputs are the control inputs. The control input bits produce one active output line. The decoder thus converts binary numbers or codes to its decimal equivalent.

The most commonly used decoder is BCD-to-decimal decoder, the circuit of which is shown below:

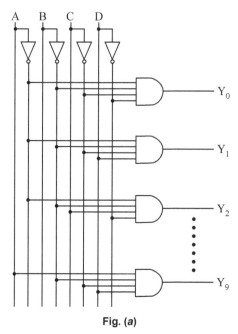

Fig. (a)

BCD is an abbreviation for binary-coded decimal. The BCD code expresses each digit in decimal number by its nibble equivalent. When computers process BCD numbers, it means that the decimal numbers were changed into BCD numbers. Now to convert back to decimal from BCD form we take group of four bits and convert to its equivalent decimal digit. Here we should keep in mind that BCD digits are from 0000 to 1001. All combinations above this *i.e.* 1010 to 1111 cannot exist in the BCD code because the highest decimal digit being coded is 9.

The circuit shown in Fig. (a) above is also called 1-of-10 decoder because only 1 of the 10 output lines is high. For *e.g.* when ABCD is 0010 ($\overline{A}\,\overline{B}C\overline{D}$), only Y_2 AND gate has all high inputs, therefore only Y_2 output is high. If ABCD changes to 1001 ($A\overline{B}\,\overline{C}D$), only Y_9 AND gate has all high inputs, thus only Y_9 output goes high.

Similarly if we check the other ABCD possibilities (0000 to 1001), we will find that the subscript of the high output always equals the decimal equivalent of the input BCD digit. Here we find that the control inputs ABCD act as the input which is being converted to its decimal equivalent. For this reason, the circuit is also called a BCD-to-decimal converter.

TTL IC7445 is used as BCD-to-decimal decoder. The pin diagram of 7445 is shown below. Pin number 16 connects to the supply voltage V_{CC} and pin 8 is grounded. Pins 12 to 15 are for BCD input (ABCD), pins 1 to 7 and 9 to 11 are for the outputs. This IC is functionally equivalent to the circuit shown in fig. (*a*) above, except that the active output line is in the low state. All the other output lines are high as shown in the truth table below:

Y_0 [1]   [16]V_{CC}
Y_1 [2]   [15]D
Y_2 [3]   [14]C
Y_3 [4]   7445   [13]B
Y_4 [5]   [12]A
Y_5 [6]   [11] Y_9
Y_6 [7]   [10] Y_8
GND [8]   [9] Y_7

Pin diagram of 7445

No	A	B	C	D	Y_0	Y_1	Y_2	Y_3	Y_4	Y_5	Y_6	Y_7	Y_8	Y_9
0	L	L	L	L	L	H	H	H	H	H	H	H	H	H
1	L	L	L	H	H	L	H	H	H	H	H	H	H	H
2	L	L	H	L	H	H	L	H	H	H	H	H	H	H
3	L	L	H	H	H	H	H	L	H	H	H	H	H	H
4	L	H	L	L	H	H	H	H	L	H	H	H	H	H
5	L	H	L	H	H	H	H	H	H	L	H	H	H	H
6	L	H	H	L	H	H	H	H	H	H	L	H	H	H
7	L	H	H	H	H	H	H	H	H	H	H	L	H	H
8	H	L	L	L	H	H	H	H	H	H	H	H	L	H
9	H	L	L	H	H	H	H	H	H	H	H	H	H	L

Truth table

Also for an invalid BCD input (1010 to 1111), all output lines will be in high state.

The logic diagram according to the truth table is shown below:

The bubble in the output lines indicates active low signal.

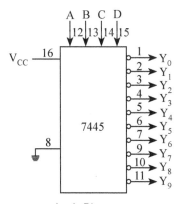

Logic Diagram

Q.11. Show how using a 3-to-8 decoder and multi-input OR gates following Boolean expressions an be realised simultaneously

$$F_1(A, B, C) = \Sigma m(0, 4, 6); \quad F_2(A, B, C) = \Sigma m(0, 5);$$
$$F_3(A, B, C) = \Sigma m(1, 2, 3, 7).$$

Ans. Since the functions are 3 variable functions and using 3-to-8 decoder we get all the 8 minterms at the decoder output, we shall use them as shown below to get the required Boolean functions.

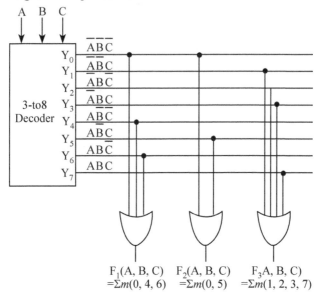

$$F_1(A, B, C) \quad F_2(A, B, C) \quad F_3A, B, C)$$
$$=\Sigma m(0, 4, 6) \quad =\Sigma m(0, 5) \quad =\Sigma m(1, 2, 3, 7)$$

Q.12. (*a*) Construct a 4 × 16 decoder using two 3 × 8 decoder.

(*b*) Explain how a decoder can work as a demultiplexer.

Ans. (*a*) IC74138 is 3 × 8 decoder IC. Using two IC74138 we can construct 4 × 16 decoder. This is made possible by using Enable input terminal. Figure below shows 4 × 16 decoder using two 3 × 8 decoder.

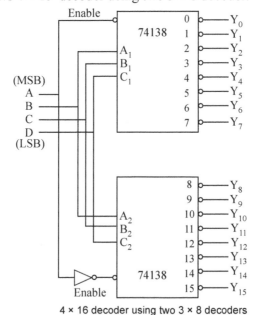

4 × 16 decoder using two 3 × 8 decoders

This decoder has four inputs A, B, C, and D. The MSB input A is connected to enable input of IC(1) and also connected after complementing to enable input of IC(2). The upper IC is enabled when input A goes low and at the same time lower IC is disabled. On being enabled the upper IC is driven by B, C and D and only one of the output lines in the upper IC goes low depending on the state of B, C and D input.

Now when A goes 1 or high, the lower IC is enabled and the upper IC is disabled. Depending on the states of B, C and D one of the output lines in the lower IC goes low. In this way the above combination of two 3×8 decoders works as a 4×16 line decoder.

(b) A decoder is similar to demultiplexer with only one exception that there is no data input. Hence if we want to use a decoder as a demultiplexer all we have to do is add a DATA input. For e.g. IC74154 is a 1-to-16 demultiplexer. It can also be used as decoder by simply grounding the DATA input. The 74154 is therefore called decoder-demultiplexer, as it can be used both as decoder and a demultiplexer.

Q.13. Explain the use of decoder/driver IC for seven-segment display of both common anode and common cathode type.

Ans. A seven segment decoder-driver is an IC decoder that can be used to drive a seven-segment indicator. There are two types of decoder-drivers corresponding to the common-code and common cathode indicators. Each decoder driver has 4 input pins (the BCD input) and 7 output pins (the a through g segments).

Figure (a) below shows the IC7446 driving a common anode indicator. Logic circuits inside the 7446 convert the BCD input to the required output. For e.g. if BCD input is 0111, the internal logic (not shown) of the 7446 will force LEDs a, b, and c to conduct. As a result, digit 7 will appear on the seven-segment indicator.

The current-limiting resistors connected externally between the seven-segment indicator and the 7446 of the Fig. (a) is to limit the current in each segment to a safe value between 1 and 50 mA, depending on how bright we want the display to be.

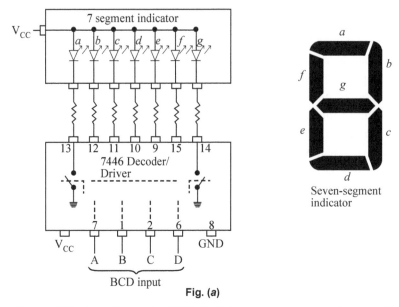

Fig. (a)

Figure (b) below shows the IC7448 driving a common cathode indicator. Again, the internal logic converts the BCD input to the required output. For example, when a BCD input of 0100 is used, the internal logic forces LEDs b, c, f and g to conduct. Th seven segment indicator then displays 4. Unlike 7446 that requires external current-limiting resistors, the 7448 has its own current limiting resistors on the chip which is connected to V_{CC}. A switch symbol is used to illustrate operation of the 7446 and 7448 in Fig. (a) and Fig. (b). Switching in the actual IC is accomplished using bipolar junction transistors (BJTs).

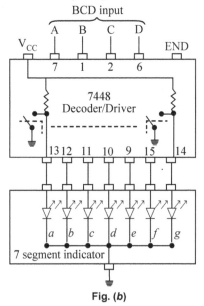

Fig. (b)

Q.14. (*a*) What is an encoder? Describe the working of decimal to BCD encoder.

(*b*) Draw the pin diagram, logic diagram and write the truth table of IC74147. Explain why IC74147 is called a priority encoder.

Ans. (*a*) An encoder converts an active input signal (decimal data) to a coded output signal. An encoder has a number of inputs, only one of which is high at a time and an m-bit code is generated depending upon which of the input is excited. Encoder is similar to multiplexer. The block diagram of an encoder is shown below:

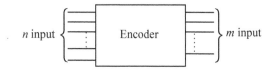

The decimal to BCD encoder is the most common type of encoder. Fig. (*a*) below shows the decimal-to-BCD encoder.

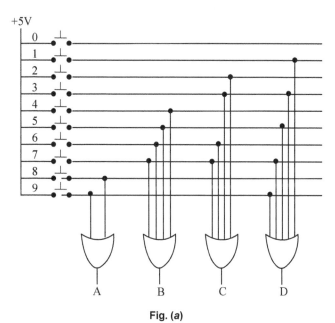

Fig. (a)

In the above circuit the switches are push-button switches like those of a pocket-calculator. When button 2 is pressed, only C OR gate has high input. Therefore in this case BCD output is ABCD = 0010. Thus decimal 2 is converted to BCD → 0010. Similarly if button 6 is pressed, B and C OR gates will have high outputs. Therefore now BCD output will be ABCD = 0110.

So, the decimal number which is needed to be converted to its equivalent BCD code, that particular switch is to be pressed and we get the required BCD output.

Ans. (*b*)

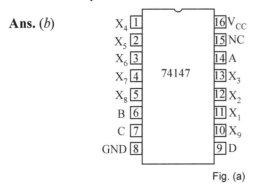

Fig. (a)

IC74147 is a decimal-to-BCD encoder. The pin diagram is shown in Fig. (a). The decimal input X_1 to X_9 connect to pins 1 to 5 and 10 to 13. The BCD output comes from pins 14, 6, 7 and 9. Pin 16 is for the supply voltage and pin 8 is grounded. The label NC on pin 15 means no connection (the pin is not used).

Logic diagram of IC 74147 is shown below:

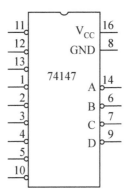

The truth table of IC 74147 is shown below:

Input									Output			
X_1	X_2	X_3	X_4	X_5	X_6	X_7	X_8	X_9	A	B	C	D
H	H	H	H	H	H	H	H	H	H	H	H	H
X	X	X	X	X	X	X	X	L	L	H	H	L
X	X	X	X	X	X	X	L	H	L	H	H	H
X	X	X	X	X	X	L	H	H	H	L	L	L
X	X	X	X	X	L	H	H	H	H	L	L	H
X	X	X	X	L	H	H	H	H	H	L	H	L
X	X	X	L	H	H	H	H	H	H	L	H	H
X	X	L	H	H	H	H	H	H	H	H	L	L
X	L	H	H	H	H	H	H	H	H	H	L	H
L	H	H	H	H	H	H	H	H	H	H	H	L

Truth table

From the truth table we find that when all inputs are high, all outputs are high. When X_9 is low, the ABCD output is LHHL which is equivalent to 9 if we complement the bits. When X_8 is only low input, ABCD is LHHH which is equivalent to 8 if the bits are complemented. If we continue in a similar manner, we can see that an active-low decimal input is being converted to a complemented BCD output.

Since it gives priority to the highest-order input as it appears from the truth table, IC74147 is called a priority encoder. If all inputs X_1 through X_9 are low, the highest of these X_9 is encoded in complemented form LHHL, giving priority to X9 over all others. Examining truth table, we can see that the highest active-low from X_9 to X_1 has priority and controls the encoding.

8

Different Types of Displays

FACTS THAT MATTER

1. **Light Emitting Diode (LED):** Light emitting diodes or LED are diodes which emits light when forward biased. The radiated emitions may be visible (red, green, yellow or orange) or invisible (infra-red) depending upon the elements used in fabricating the diode.

   When a diode (such as GaAsP and GaP discrete light emitting diodes) is forward biased, free electrons cross over the junction and are trapped in the holes. While this recombination takes place, the free electrons jump from their high energy level to a lower one and as a result some energy is radiated out in the form of light in case of LED. By using elements such as Gallium, Arsenic and Phosphorus the colour of the radiated energy can be red, green, yellow, orange and infra-red. Such LEDs are used in digital instruments, calculators, watches etc. The infra-red LEDs are used in burglar-alarm and other devices requiring invisible light.

2. **Liquid Crystal Display (LCD):** LCDs as used in watches are twisted nematic field effect devices. Liquid crystals are sandwiched between front and back planes in thin glass to make a plastic or glass case. Some characteristic features of liquid crystal displays are:

   (*i*) To align liquid crystal molecules, transparent conductor patterns are coated inside glass planes with a special chemical film.

   (*ii*) Liquid crystals have the property intermediate to liquids and solid with a finite range of temperature.

(*iii*) They require low voltage and current, thus can be driven directly by low current CMOS circuits. They have a light sensing capability to increase the intensity of the display in bright ambient light, while reducing it in dim light. Liquid crystals without plastic polarizers are available.

Dynamic scattering display are also available. These liquid crystals consist of two glass plates with liquid crystal in between. Normally the back plate is coated with a thin, transparent layer of conductive material. The front plate has a photo etched conductive coating, such as seven segment pattern. When no power is applied to the plates, the molecules of the liquid crystal align themselves either parallel to the glass plates or perpendicular and appear transparent. But when a voltage is applied it causes the molecules to scatter and the liquid crystal then reflects ambient light and appears milky white. A dynamic scattering display like this produces light character on a dark background.

3. **Encoder:** An encoder is a device which converts alphanumeric characters to binary numbers.

4. **Decoder:** A decoder converts binary words in to alphanumeric character.

5. **Digital display:** There are three types of digital displays:
   (*i*) Discrete display
   (*ii*) Bar-matrix display
   (*iii*) Dot-matrix display

OBJECTIVE TYPE QUESTIONS

1. DC forward voltage is needed to emit light in case of LED or LCD?
2. When all the seven segments of a display are energized, the number shown will be _____?
3. Out of LCD and LED which display consumes the least power?
4. When the input to a seven segment decoder is 0101, the number on display will be _____.
5. The segments of a seven segment display are lettered to a _____.
6. Current drawn when the number 8 is on an LED display is _____.
7. Current supplied to four digit LCD what reads 8888 is of the order of 560 μA or 560 nA.
8. Dot matrix display can display decimal digit, alphabets and also other symbols (True r False).
9. Which of the following methods cannot be used for displaying the alphanumeric?

(*a*) Discrete method, (*b*) Bar-matrix method, (*c*) ASCII code, (*d*) Dot-matrix method.

10. Three ways to display alphanumerics are _____, _____ and _____.

Answers

1. LED.	**2.** 8.
3. LCD.	**4.** 5.
5. Clockwise direction.	**6.** 140 mA
7. 560 nA.	**8.** True.
9. ASCII code,	

10. Discrete display, Bar-matrix display and Dot-matrix display.

SHORT ANSWER TYPE QUESTIONS

Q.1. What are the major advantages of LEDs in electronic display?

Ans. The major advantages of LEDs in electronic display are as follows:

(*i*) LEDs are available which emit light in different colours like red, green, yellow or orange.

(*ii*) LEDs are miniature in size and they can be stacked together to form numeric and alphanumeric displays in high density matrix.

(*iii*) LEDs area manufactured with same type of technology as is used for transistors and ICs and they are economical and have a high degree of reliability.

(*iv*) The light output from an LED is a function of the current flowing through it. Hence intensity of light emitted from LEDs can be smoothly controlled.

(*v*) LEDs have a high efficiency as emitters of electromagnetic radiation. They require moderate power for their operation.

(*vi*) The switching time (both for ON and OFF) is less than Ins and therefore, they are useful where dynamic operation of large number of arrays is involved.

(*vii*) LEDs are rugged and can therefore withstand shocks and vibrations.

(*viii*) LEDs can be operated over a wide range of temperature.

(*ix*) LEDs have a long life about 1,00,000 hours.

(*x*) LEDs can be directly driven from diode-transistor logic and transistor logic.

Q.2. What are the disadvantages of LED?

Ans. (*i*) LEDs require high power compared to LCD.

(*ii*) LEDs are not suited for large area displays, primarily because of their high cost.

Q.3. What are the advantages and disadvantages of LCD?

Ans. The advantages of LCD are:

    (*i*) LCDs have low power consumption. A seven segment display requires about 140 μW compared to 40 mW required by LEDs.

   (*ii*) They have a low cost.

  (*iii*) LCD can be driven directly from IC chips. Driver circuits are not required.

    The disadvantages of LCD are:

    (*i*) LCDs are very slow devices. The turn-on and turn off times are quite large, typically turn-on requires a few ms and turn-off is 10 ms.

   (*ii*) Poor visibility under low ambient lighting.

  (*iii*) When used on d.c., the life span of LCDs are quite small. Hence LCDs are used with a.c. supplies having a frequency less than 50 Hz.

  (*iv*) They occupy a large area.

   (*v*) LCDs can be operated over a limited temperature range.

  (*vi*) Due to chemical degradation the life time is around 50,000 hours.

 (*vii*) LCDs have comparatively slower response rate time factor.

LONG ANSWER TYPE QUESTIONS

Q.1. Differentiate between LED and LCD.

Ans. The difference between LED and LCD are as follows:

LED	LCD
(*i*) LEDs consume more power.	(*i*) LCDs consume very less power.
(*ii*) Due to high power requirement, LED requires external interface circuits (called LED driver circuit) when driven from ICs.	(*ii*) LCD can be driven directly from IC chips. Driver circuits are not re-quired.
(*iii*) The brightness level is very good for LEDs.	(*iii*) LCDs have moderate brightness level.
(*iv*) Commercially available LEDs have operating temperature range of –40 to 85 degree celcius.	(*iv*) Comparatively less temperature limit –20 to 60 degree celcius.
(*v*) Lifetime of LED is around 1,00,000 hours.	(*v*) Due to chemical degradation lifetime is around 50,000 hours.
(*vi*) LEDs have wide viewing angle.	(*vi*) LCDs have limited viewing angle.
(*vii*) Operating voltage range for LED is 1.5 V to 5V DC.	(*vii*) Operating voltage range for LCD is 3 V to 20 V DC.
(*viii*) LED displays have better colour accuracy.	(*viii*) LCDs have less colour accuracy.
(*ix*) LED displays have better contrast and more detailed image in the dark areas compared to LCDs.	(*ix*) LCD displays are not able to create the dark areas of the images on the screen which leads to loss of image quality.

(Contd...)

LED	LCD
(x) LED displays are environment friendly since mercury is not been used in it.	(x) The back lights of LCD display contain mercury which poses a risk of environment. Due to broken back lights in LCD panels, mercury vapour is released in environment which is hazardous to health.
(xi) LED displays are equipped with a faster response rate time factor.	(xi) LCDs have comparatively slower response rate time factor.
(xii) LED displays are more expensive than LCDs.	(xii) LCDs are less expensive than LED displays.

Q.2. Explain the different methods for alphanumeric displays.

Ans. To most of the users of digital systems various numbers such as binary or BCD numbers do not convey the informations directly. For this reason, we make use of electronic circuits to decode such numbers to decimal numbers. In some applications beside decimal numbers, the use of alphabet letters for display is also essential. There are three ways in which alphanumeric characters are displayed:

(i) Discrete method → Display is 1, 2 and so on.

(ii) Bar-matric method → Display is !, 2 and so on.

(iii) Dot-matrix method → Display is 1, 2 and so on.

(i) **Discrete Method:** In this method, a single light source produces each character. For this purpose we make use of Nixie tube as shows below:

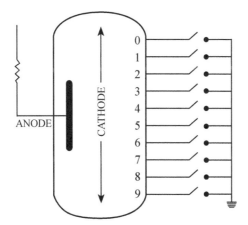

Nixie table has one anode and several cathodes. Each cathode is shaped like the character to be displayed. In normal situation *i.e.* when they are not activated, these characters are transparent. When a particular or desired cathode is grounded, it causes the neon gas around that cathode to ionize and it results in a glow and the particular character (corresponding to the cathode grounded) is displayed.

For e.g. if we want 6 to be displayed then the corresponding switch is closed so that, that cathode is grounded, and neon gas will ionize around that cathode and so 6 will be displayed.

In discrete method we use one light source for the display of each symbol. This requires activating only one pin on the display device to produce each symbol.

(*ii*) **Bar-matrix method:** In bar-matrix method, one or more than one light sources shaped like bars or segments are involved in the display of a symbol. To produce each symbol it is necessary to activate particular one or more pins on the display device. The most common bar-matrix device is the seven segment indicator *i.e.* seven LEDs labeled a through g as shown below: (Fig (a)). Each LED is shaped like bar.

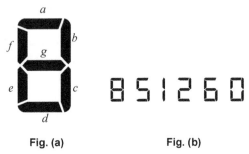

| Fig. (a) | Fig. (b) |

By forward biasing different LEDs, we can display the digitals 0 through 9 (some of them are shown in Fig. (b). For instance, to display 8, we need to light up all the segments *a* through *g*. To light up 5, we need segments *a*, *c*, *d*, *f* and *g*.

Seven segment indicators may be the common-anode type where all anodes are connected together (Fig. (*a*) below), or the common cathode type where all the cathodes are connected together (Fig. (*b*).

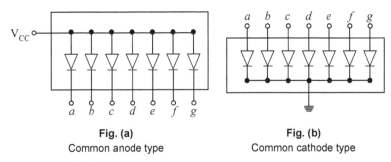

| Fig. (a) | Fig. (b) |
| Common anode type | Common cathode type |

With common anode type we have to connect a current limiting resistor between each LED and ground. The size of this resistor determines how much current flows through the LEDs. It should be between 1 to 50 mA. The common cathode type (Fig. b) uses a current limiting resistor between each LED and +V_{CC}. A seven segment decoder driver is an IC decoder that can be used to drive a seven-segment indicator. There are two types of decoder-drivers corresponding to the common-anode and common-cathode indicators. Each decoder-driver has 4 input pins (the BCD input) and 7 output pins (*a* through *g* segments).

The logic circuits inside the IC 7446 (which drives common-anode indicator) convert the BCD input to the required output. For *e.g.* if the BCD input is 0110, the internal logic of IC7446 will force LEDs *a*, *c*, *d*, *e*, *g* and *f* to conduct. As a result 6 will appear on the seven-segment indicator. The current limiting resistors between the seven-segment indicator and 7446 will limit the current in each segment to a safe value between 1 to 50 mA.

Similarly IC7448 drives a common cathode seven-segment indicator.

(*iii*) **Dot-marix method:** In this method devices called dot-matrix devices are used. Dot-matrix devices have many individual light sources shaped like dots. As an example a typical 5 × 7 LED matrix is shown below: For activating an LED in this matrix cathode is to be grounded and a voltage is to be applied to its anode.

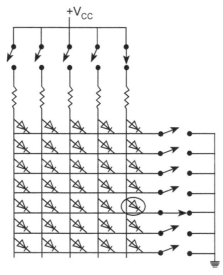

In the Figure the circled LED is lit because a voltage is applied to the fifth vertical column and a ground is applied to the fifth horizontal row. However by applying voltages to more than one column and grounding more than one row, we can display any decimal digit or alphanumeric character.

For displaying a character with dot matrix readout, the desired LEDs are not activated simultaneously rather they are activated a row at a time in rapid succession. When this process is repealed fast enough an alphanumeric character is displayed without any flicker. The display by this method is like: 1 2 3.

9

Digital Logic Families

FACTS THAT MATTER

1. **Logic Family:** Many complex digital functions have been realized in a variety of forms and each form is referred to as a logic family. Basically there are two types of semiconductor devices: Bipolar and Unipolar. Digital ICs are fabricated by employing either the bipolar or the unipolar technologies and are accordingly referred to as bipolar logic family or unipolar logic family.

2. **Bipolar logic families:** The main elements of a bipolar IC are resistors, diodes, capacitors and transistors. Basically there are two types of operations in bipolar ICs:

   (*i*) Saturated and (*ii*) non-saturated

   In saturated logic, the transistors n the IC are driven to saturation whereas in case of non-saturated logic, the transistors are not driven into saturation.

   The saturated bipolar logic families are:

   (*i*) Resistor-transistor Logic (RTL).

   (*ii*) Integrated-injection Logic (I^2L).

   (*iii*) Diode-transistor Logic (DTL)

   (*iv*) High-threshold Logic (HTL) and

   (*v*) Transistor-transistor Logic (TTL)

   The non-saturated bipolar logic families are:

   (*i*) Schottky TTL.

   (*ii*) Emitter-coupled Logic (ECL).

3. **Unipolar logic families:** MOS devices are unipolar devices and only MOSFETs are employed n MOS logic circuits. The MOS logic families are:

(*i*) PMOS

(*ii*) NMOS and

(*iii*) CMOS

While in PMOS only *p*-channel MOSFETs are used and in NMOS only *n*-channel MOSFETs are used, in complementary MOS (CMOS) both *p*- and *n*-channel MOSFETs are employed and are fabricated on the same silicon chip.

4. **Classification:** Digital ICs are classified as per level of integration as follows.

(*i*) Small scale integrated circuits (SSI)

(*ii*) Medium Scale Integrated Circuits (MSI)

(*iii*) Large Scale Integrated Circuits (LSI)

(*iv*) Very Large Scale Integrated Circuits (VLSI)

5. **Definitions:**

(*i*) **SSI:** Those ICs which have total number of gates less than 12 integrated on the same chip are called SSI (Small Scale Integration).

(*ii*) **MSI:** Those ICs which have 12 to 100 gates integrated on the same chip are called MSI (Medium Scale Integration)

(*iii*) **LSI:** Those ICs which have 100 or more but less than 1000 gates integrated on the same chip are called LST (Large Scale Integration).

(*iv*) **VLSI:** Those ICs which have 1000 or more gates integrated on the same chip are called VLSI (Very Large Scale Integration).

6. **Characteristics of digital ICs:** The various characterisation of digital ICs used to compare their performances are:

(*i*) Propagation delay

(*ii*) Speed of operation

(*iii*) Noise margin

(*iv*) Fan in and fan out

(*v*) Power dissipation

(*vi*) Logic levels

(*vii*) Power supply requirements

(*viii*) Operating temperature.

(*i*) **Propagation delay:** The signals through a gate take a certain amount of time to propagate from the input to the output. This interval of time is defined as the propagation delay of the gates. Propagation delay is measured in nano seconds (ns).

(*ii*) **Speed of operation:** The speed of a digital circuit is specified in terms of the propagation delay time. The higher-speed circuit is one that has a smaller propagation delay.

(*iii*) **Noise margin:** It is the maximum noise voltage added to an input signal of a digital circuit that does not cause an undesirable change in the circuit output. In other words a measure of noise immunity of a circuit is called the noise margin. It is expressed in volts and represents the maximum noise signal that can be tolerated by the gate. Noise signals can be AC or DC. There are two values of noise margin specified for a given logic circuit:

V_{NH} (High level Noise Margin)

V_{NL} (Low level Noise Margin)

so that

$V_{NH} = V_{OH} (min) - V_{IH} (min)$

$V_{NL} = V_{IL} (max) - V_{OL} (max)$

Here V_{OH} (min) is the lowest possible HIGH output from driving gate and V_{IH} is the lowest possible HIGH input that the load gate can tolerate. Similarly noise margin V_{NL} is the difference between the maximum possible low input that a gate can tolerate V_{IL} and he maximum possible low output of the driving ate V_{OL}.

(*iv*) **Fan in and Fan out:** For a logic circuit, fan-in represents the number of inputs that can be connected to the logic circuit input. The Fan-out specifies the number of unit inputs that can be driven by a logic element.

In a logic circuit, normally the output of one gate is connected to the input of some other gate. Each input requires a certain amount of power from the gate output and each subsequent addition of a gate adds to the load of the gate. It is very important for the smooth working of the circuit to define its fan-out as this is the maximum number of inputs that can be connected to the output of a gate without making it over loaded. Sometimes when it is necessary to connect more number of gates than its specified fan-out we make use of buffer gates to avoid malfunction and overloading of the gate. High fan out is advantageous because it reduces the need for buffer gates.

(*v*) **Power dissipation:** The power dissipation for a logic circuit is defined as the power required for the circuit to operate with 50% duty cycle at a particular frequency. The normal working power required for a gate to operate ranges from few microwatts to a few milliwatts per gate. For having less power dissipation in the circuit, the number of gates used should be minimized. If I_{CC} is the current that it draws from the supply V_{CC}, power dissipation is given by $V_{CC} \times I_{CC}$.

(*vi*) **Logic levels:** The specifications of voltages corresponding to logic 0 and 1 are of great importance as these effect power dissipation, noise margin and operating speed. Normally in a circuit, we use the logic families which have same logic voltage level. If we use ICs of different families their logic voltage levels may be different and we have to use interfacing device before connecting such two ICs.

(*vii*) **Power supply requirement:** The supply voltage(s) and the amount of power required by an IC are important characteristics required to choose the proper power supply.

(*viii*) **Operating temperature:** All the logic gates are made of ICs making us of semiconductor devices, which are temperature sensitive. The temperature range in which an IC functions properly must be known. The accepted temperature ranges are 0 to +70°C for consumer and industrial applications and –55%c to +125°C for military purposes.

7. **Resistor transistor logic (RTL):** RTL was the first family of logic circuits which was in common use before the development of ICs. RTL circuits consist of resistors and transistors. In this family, transistors are connected in series or parallel with resistors to perform logic operations. The transistor used, work as ordinary switches. A resistor transistor logic NOR gate in positive logic is shown in figure below:

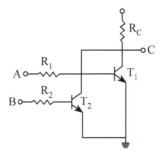

The inputs are A and B and the output is C. When input A to transistor T_1 is high (5 V), T_1 will conduct and make the output low (0). Similarly when input B to transistor T_2 is high, T_2 will conduct and make the output low. When both inputs are low, the transistor T_1 and T_2 will be cut off and the output will be high. When both A and B are high, T_1 and T_2 will conduct, so output will be low. So the truth table for the above circuit is

A	B	C
0	0	1
0	1	0
1	0	0
1	1	0

From this we conclude that this circuit performs the NOR logic operation for RTL logic circuits.

The propagation delay for RTL logic circuits is ~12 n sec, noise margin ~0.2 V, power dissipation is ~30–100 mW, fan out is 4 and power supply voltage V_{CC} is 5 V.

RTL circuit can also be connected such that it can work as NAND gate as shown below:

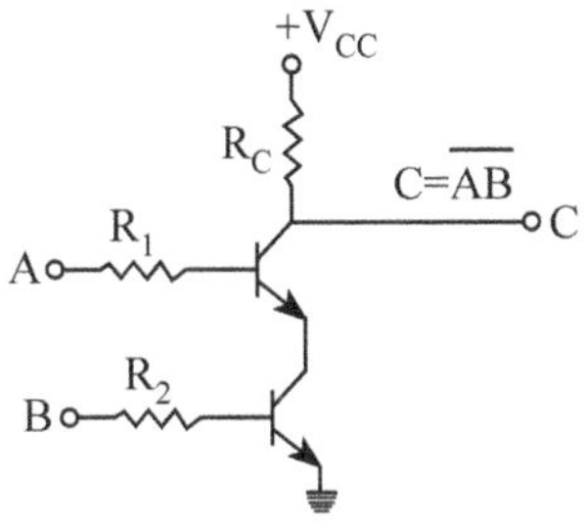

A	B	C
0	0	1
0	1	1
1	0	1
1	1	0

The RTL circuit has the disadvantage due to the presence of resistances R_1 and R_2 which causes a reduction in the switching speed. To increase this speed a capacitor C each in shunt with R_1 and R_2 can be connected. This capacitor allows rapid charging during the turn-ON period of transistors. This improved version of the gates is named as RCTL gates.

RTL technology described above is not used frequently but RTL inverter is used in bipolar ICs. An RTL inverter gate is shown below:

$$+V_{CC}$$

R_C

$C = \overline{A}$

Ao—ww—

8. **Diode transistor logic (DTL):** The diode-transistor logic is somewhat more complex than RTL but because of its greater fan-out and improved noise margins it has replaced RTL. Diode-transistor logic (DTL) makes use of diode AND or OR gate and transistor inverter i series for making it to be used as NAND or NOR logic operator. Figure (*a*) below shows a two input NAND gate using DTL.

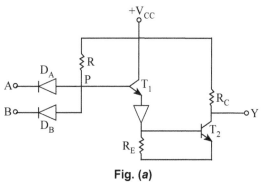

Fig. (a)

The diodes D_A and D_B are input diodes. Input diodes conduct through the resistor R, if the corresponding input is in low state. Corresponding to high state input, the diodes are non-conducting. Therefore if input A is low, diode D_A conducts reducing the potential of point P to low value. Transistor T_1 and T_2 are cut-off and output is high or 1. A similar action occurs if input B is low or if both A and B are low. On the other hand if both inputs are high, the input diodes D_A and D_B will not conduct and consequently the current flowing from V_{CC} through R drives transistor T_1 and T_2 into saturation. Therefore the output Y of T_2 is $V_{CC(sat)}$ or low or 0.

Thus this DTL circuit is acting as NAND gate. The DTL logic family has a power dissipation of 60 mW, propagation delay of 30 n sec, noise margin of 0.7 V, fan-out of 8 and power supply voltage of 5 V.

9. **High threshold logic (HTL):** In industrial environment due to presence of a number of electric motors, on-off control circuits, high voltage switches etc; the noise level is very high and the logic families like RTL, DTL do not perform properly rather there is malfunctioning. In order to take into consideration of higher level of noise a DTL gate is designed with a provision of higher power voltage (15 V instead

of 5 V). The normal diode D is replaced by a zener diode having zener breakdown voltage of 6.9 V and the resistances of the circuit are so modified to obtain almost same current as in DTL. A HTL two input NAND gate is shown in Fig. (a) below. Due to large resistance values, the propagation delay time is increased. It is almost 100 ns. The sensitivity to temperature of HTL gate is less than that of DTL.

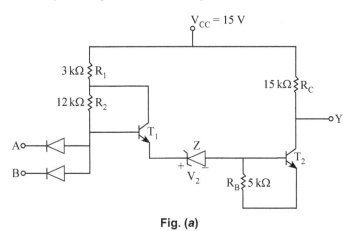

Fig. (a)

10. **Integrated injection logic (I_2L or IIL):** This type of logic (I^2L) is not used in single gates as was the case for other logic families, rather it is more useful in the ICs where thousands of gates are made on one chip such as microprocessor and digital watch circuit. This is a new class of bipolar saturated logic circuit I^2L does not contain resistors, hence they are smaller in size and specially used in very large scale integrated circuits.

Figure below shows two input integrated injection logic NOR circuit.

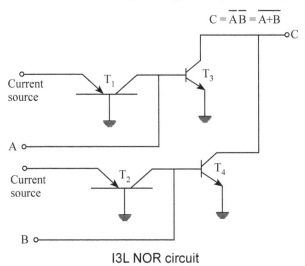

I3L NOR circuit

The transistors T_1 and T_2 act as current source feeding current to the bas of the transistor T_3 and T_4. Now if input A is low, the current source of transistor T_1 is shortened to the grounds and T_3 does not conduct *i.e.* it is off. If in this case B input is high, the output will be low. So output will be low when either of the input A or B or both the inputs A and B are high. The output will be high only when both the inputs A and B are low. Hence this gate performs the function of NOR gate. The collector of transistor T1 and the base of transistor T_3 can be the same element and similarly, the collector of T_2 and base of T_4 are the same element. This property further reduces the size of the gate.

The simplicity and their low power enables us to make many gates in a small chip. Also as the current is constant (as there are no voltage drops) no switching transients are produced on the line. By manipulating injector current, the propagation delay and the power dissipation can be varied over a large range. The other advantage of I^2L is that it can easily be formed on the same chip with bipolar analog circuits.

11. **Transistor transistor logic (TTL):** TTL is the most popular amongst all logic families. In case of DTL, we had the problem of slow speed due to charge stored in the base of second transistor. This speed limitation in DTL is overcome in TTL. These types of logic circuits make use of transistor with multiple emitters and the inputs are connected to the emitter. The base gate of this family is a NAND gate and circuit of a basic TTL NAND gate is shown below in Fig. (*a*). This basic gate has one main limitation of the gate speed and is due to the effects of capacitance which appears across the output of the gate from the collector T_2 to ground. This capacitance is made up of capacitance of the output transistor

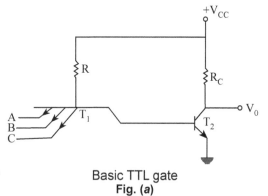

Basic TTL gate
Fig. (a)

itself. When T_2 is cutoff, this output capacitance must charge from V_{CC} through passive pull-up resistor R_C. The output rises from logic 0 to logic 1 within time $R_C C_0$ (time constant). This time can be reduced by reducing R_C. But this reduction will increase power dissipation in R_C and T_2 (while it is conducting) and also it makes it more difficult to saturate T_2. To overcome this problem we make use of active pull up or Totem-pole output circuit arrangement. In this arrangement R_C is replaced with an active device T_3 and diode D_3 as shown in Fig. (*b*) below:

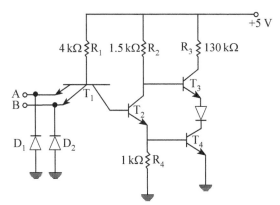

Basic TTL two input NAND gate

Fig. (b)

The main difference between TTL and other logic circuits is that in TTL we make use of multiple emitter transistor at the input circuit and the output stage normally contains two transistors. All the unused inputs of TTL gates may add to the noise hence they should be connected to a used input or to V_{CC} through a resistor. An input left open will act as if it is connected to high level as there is no base emitter current path for transistor T_1 and may pick up the noise pulse.

The output configuration as shown in Fig (b) above is an active pull-up or toten-pole output i.e. out of T_3 and T_4 when one transistor is ON the other is OFF.

We can also have two more types of output arrangements in TTL circuits called open collector arrangement and tri-state or three-state output.

Open Collector Outputs: In case of toten-pole output stage, for LOW output the lower transistor T_4 is ON and the upper transistor T_3 is OFF and vice-versa for a HIGH output. In case of open collector output the upper transistor is removed as shown in Fig. (c) below. Now the output will be a LOW when T_4 is ON and the output will float (not HIGH or LOW) when T_4 is OFF. This implies that an open-collector (OC) output can sink current, but it cannot source current.

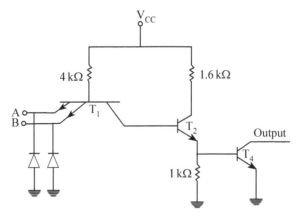

TTL NAND gate with OC output

Fig. (c)

In order to get an open collector output to produce a HIGH an external resistance called a pull-up resistor should be used as shown in Fig. (*d*). Under the situation when T_4 is OFF (open) the output is nearly 5 V (HIGH) and when T_4 is ON (short) the output is approximately 0 V (or LOW). The maximum value for the pull-up resistor depends on the size of the outputs load and the leakage current through T_4 (I_{OH}) when it is OFF. In general the value of pull-up resistor is 10 kΩ. This value is neither too small to allow excessive current flow when T_4 is ON, nor it is too large to cause an excessive voltage drop across itself when T_4 is OFF.

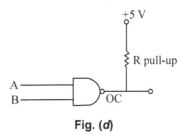

Fig. (*d*)

Tristate TTL logic gate or 3 state TTL gate: As ordinary totem-pole device cannot be wire-AND due to excessive power dissipation so we use a new type of totem-pole device called tri-state TTL. It offers high switching speed operation of the totem-pole device and also allows the outputs of two different gates to be wire-ANDed as this permits three possible output states *i.e.* HIGH, LOW and HIGH impedance. Such circuit is shown below in Fig. (*e*) and its symbol is shown in Fig. (*f*).

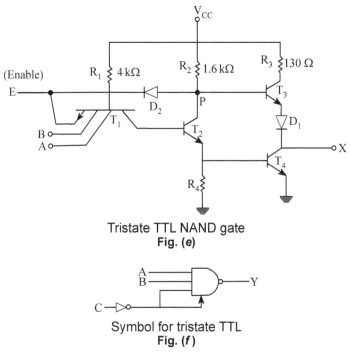

Tristate TTL NAND gate
Fig. (e)

Symbol for tristate TTL
Fig. (f)

Such gates can bes directly connected to a common line (bus). Only that gate gets connected which has its enable HIGH. All other enables are kept at low. When the enable is in disabled state (LOW) both the transistors T_3 and T_4 remain OFF. As a result the output is neither connected to V_{CC} or ground, it is isolated whatever may be the inputs. Under the situation the gate is said to be in high impedance (Hi-Z) state. We can also say the output is in open or floating state. If enable terminal E is held HIGH, the diode D_2 does not allow this high voltage at E to reach P. The voltage at P depends upon the inputs as it happens in case of ordinary gate without tristate circuit. With enable (E) HIGH, X is LOW only if both the inputs are high; else X is HIGH. In this way gate has three possible states LOW, HIGH and Hi-Z. The tristate logic is used in multiplexing/demultiplexing, biodirectional bus configuration in microprocessors and in expandable memory systems.

The TTL logic gates are faster having propagation delay of 15 *ns* and has high power dissipation of 100 mW. The noise margin is 0.4 volts and fan out is 10. The power supply required is 5 V.

12. **Non-saturated bipolar logic families:**

   (*i*) **Schottky TTL.** By using a special type of diode clamp between base and collector, the schottky TTL removes the storage time of the transistors by preventing them from going into saturation state. This increases the speed of operation without an excessive increase in power dissipation. The special type of diode is called schottky diode and is made with the junction of a metal and semiconductor.

A schottky diode works faster as compared with conventional diode because the electrons which have crossed the junction and entered the metal when the current is flowing are similar to the conduction electrons of the metal. As these electrons are majority carriers, there is no delay in the recombination with minority carrier as happens in case of semiconducor diodes. Due to use of aluminium or platinum silicides, the forward drop in case of schotky diode is less than that for an ordinary diode. This schottky diode has forward drop of 0.3 V and has no problem of storage time and can be switched on or off very fast. Such a diode connected from base to the collector switches very fast as the transistor is not allowed to saturate. The symbol used for a schottky-diode clamped transistor is

The circuit for a schottky NAND gate is shown in Fig. (a) below. The transistors used in this circuit are schottky clamped transistors so the storage delay or saturation delay is virtually eliminated which results in better switching time.

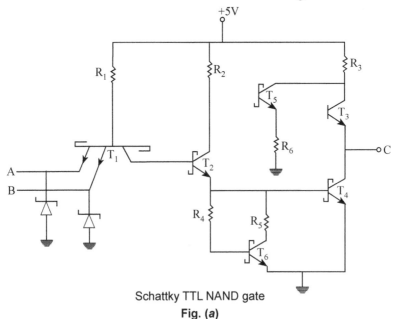

Schattky TTL NAND gate

Fig. (a)

The working of this circuit is more or less same as that of TTL NAND gate except that in this case transistor does not saturate. The members of this logic family use less power as compared to high power TTL yet they are two times faster than high power TTL and almost 4 times of standard TTL. These devices are designated as 74S00/54S00.

Low power Schottky: If we increase the values of resistance of circuit for Schottky TTL gate, these circuits are converted into low power Schottky TTL. This is a family which has minimum propagation delay, fast switching action and low power dissipation. These families are represented by 74LS00/54LS00. These require only one fifth of the power required by standard TTL and also has faster speed than standard TTL. A low power schottky gate has a power dissipation of 2 mW and a propagation delay of 10 ns. In practice the low power schottky and standard TTL are mostly used by the designer in digital electronics.

(*ii*) **Emitter coupled logic (ECL):** These ECL logic circuits are also called current mode logic (CML) or current steering logic or non-saturated logic. ECL is the fastest of all logic families and therefore is used in applications where very high speed is essential. High speeds have become possible in ECL because the transistors are used in difference amplifier configuration in which they are never driven into saturation and thereby the storage time is eliminated. Here rather than switching the transistors from ON to OFF and vice-versa, they are switched between cut-off and active regions. Propagation delays of less than 1 ns per gate have become possible in ECL.

Basically ECL is realised using difference amplifier in which the emitters of two transistors are connected and hence it is referred to as emitter-coupled logic.

ECL are not very popular owing to their high cost over TTL, difficulty in cooling, difficulty in inter-connecting. The symbol of an ECL OR/NOR gate is

13. **MOS circuits (Unipolar):** In large scale integrated circuits we make use of metal oxide semiconductor field effect transistors (MOSFET). Through these devices are slow, delicate and lack drive characteristics but because of the easy manufacturing processes, small size and low power dissipation these MOSFETs are widely used in LSI circuits. MOSFETs are very popular nowadays and find wide application in

logic gates, shift registers and semiconductor memories. In this type of transistors we make use of only one type of carriers *i.e.* either electrons or holes, hence it is called unipolar device. These transistors can be either *p*-channel MOSFET (PMOS) or *n*-channel MOSFET (NMOS). It is also possible to fabricate enhancement mode *p*-channel and *n*-channel MOS devices on the same chip. Such devices are referred to as complementary MOSFETs and logic based on these devices is known as CMOS logic. The power dissipation is extremely small for CMOS and hence CMOS logic has become very popular.

CMOS logic: A complementary MOSFET (CMOS) is obtained by connecting a *p*-channel and an *n*-channel MOSFET in series with drains tied together and the output is taken at the common drain. Input is applied at the common gate formed by connecting the two gates together as shown in Fig. (*a*) below.

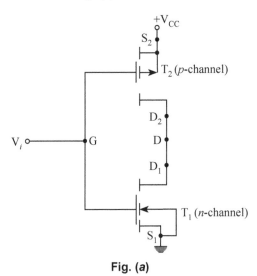

Fig. (a)

In a CMOS, *p*-channel and *n*-channel enhancement MOS devices are fabricated on the same chip which makes its fabrication more complicated and reduces the packing density. But because of negligibly small power consumption, CMOS is ideally suited for battery operated systems. Its speed is limited by substrate capacitances. To reduce the effect of these substrate capacitances, the latest technology known as silicon on sapphire (SOS) is used in microprocessor fabrication which employs an insulating substrate (sapphire). CMOS is becoming very popular in MSI and LSI areas.

The basic CMOS logic circuit is an inverter shown in Fig. (*a*) above. Here the logic levels are 0 V (logic 0) and V_{CC} (logic 1). When $V_i = V_{CC}$, T_1 turns ON and T_2 turns OFF. Therefore $V_O = 0$ V and since the transistors are connected in series the current I_D is very small. On the other hand when $V_i = 0$ V, T_1 turns OFF and T_2 turns ON giving an output $V_O = V_{CC}$ and I_D is again very small. In either logic sate T_1 or T_2 is OFF and the quiescent power dissipation which is the product of the OFF leakage current and V_{CC} is very low:

More complex functions can be realised by combination of inverters.

14. **CMOS characteristics:**

   (*i*) **Propagation Delay:** Generally propagation delay time for CMOS is higher than that of TTL devices and varies from 25 ns to 100 ns. To increase the operation speed device should be operated at higher supply voltages and load capacitance should be reduced.

   (*ii*) **Fan out:** Since MOS devices have very high input impedance, therefore, the fan-out is large. But driving a large number of MOS gates increases the capacitance at the output which reduces the speed of MOS gates.

   (*iii*) **Noise margin:** Noise margin of CMOS logic ICs is considerably higher than that of TTL ICs. These circuits are available with wide supply voltage range and the noise margin increases with the supply of voltage V_{CC}. The noise margin of CMOS is roughly 0.45 V_{DD}. If operating voltage is 12 V, the noise margin will be 5.4 V: As a result these devices are relatively immune to undesirable switching due to noise pickup.

   (*iv*) **Power supply requirement:** CMOS devices can work over a fairly large voltage range which extends from 3 V to 15 V, which is not the case with TTL devices. Power dissipation level increases with the supply voltage. So where power dissipation is important consideration CMOS devices should be operated at lower voltage. Now operating on low voltage increases propagation delay time and decreases noise immunity. So when noise margin ad propagation delay are important consideration, CMOS should be operated at high voltage preferably above 9 V.

   (*v*) **Power dissipation:** The average or static power dissipation of a CMOS device is around 10 mW. However this increases whenever there is a change from HIGH to LOW or LOW to HIGH state and the magnitude of increase depends on the frequency of operation *i.e.* with the switching speed and also on capacitive load.

   (*vi*) **Floating inputs:** As is the case with TTL devices a floating or open input is equivalent to a high input. Hence a floating input in CMOS gate is very susceptible to noise pick-up since input impedance is very high. This causes an increase in power dissipation. Owing to the drifting gate voltage, the CMOS gate may drift into a linear mode of operation. A CMOS input should not be left floating, rather it should be either connected to supply voltage terminal or one of the used inputs provided that the fan-out of the signal source does not exceed.

OBJECTIVE TYPE QUESTIONS

A. Fill in the blanks:

1. The logic family TTL having highest speed is _____.
2. The fan-out of TTL gate is about _____.
3. The _____ family has maximum fan-out capacity.
4. The logic family which dissipates maximum power is _____.
5. The _____ family has good noise immunity.
6. LSI and VISI devices use _____ technology.
7. The first logic family was _____.
8. In a CMOS device if the frequency of operation is increased, the dynamic power consumption will _____.
9. If two similar RTL gates are wire-ANDed and if each gate has a fan-out of 5, the fan-out of the combined gate will be _____.
10. The logic family having maximum fan-out is _____.
11. ICs of 5400 series belong to _____.
12. ICs of 7400 series have operation temperature range of _____.

Answers

1. Schottky TTL.	**2.** 10	**3.** CMOS.
4. ECL.	**5.** CMOS.	**6.** MOS.
7. RTL.	**8.** Increase.	**9.** Ten.
10. CMOS.	**11.** Military range.	**12.** 0° to +70°C.

B. State True or False:

1. ECL is faser than other IC technologies and it has higher fan-out than TTL.
2. MOS normally consumes less power and is slower than bipolar.
3. MOSFET can be operated in enhancement mode and depletion mode.
4. PMOS is faster than CMOS.
5. TTL and IIL families are saturated bipolar logic families.
6. Schottky TTL and ECL are non-saturated logic families.
7. PMOS, NMOS and CMOS are unipolar logic families.
8. TTL, uses field effect transistor.
9. ECL is the fastest logic family.

10. 5 V dc are required for TTL IC.

11. In ideal logic family fan in and fan out both should be high.

12. CMOS has the highest range of operating voltage.

13. Wired-logic should not be used for CMOS devices.

Answers

1. True.   **2.** True.   **3.** True.   **4.** False.   **5.** True.   **6.** True.   **7.** True.

8. False.   **9.** True.   **10.** True.   **11.** True.   **12.** True.   **13.** True.

SHORT ANSWER TYPE QUESTIONS

Q.1. Why shouldn't totem-pole output be wired together?

Ans. If we use a regular totem-pole output gates, when a gate having a HIGH output (5 V) is connected to another gate having a LOW output (0 V) it will result in direct short circuit causing either or both gates to burn out.

Q.2. How do open collector outputs differ from totem-pole outputs?

Ans.

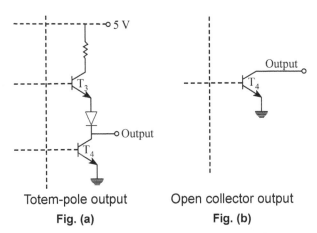

Totem-pole output   Open collector output

Fig. (a)   **Fig. (b)**

In case of totem-pole output, for LOW output the lower transistor T_4 is ON and the upper transistor T_3 is OFF and vice versa for HIGH output, whereas in case of open collector output the upper transistor is removed as shown in Fig. (*b*) above. Now the output will be a LOW when T_4 is ON and the output will float (not HIGH or LOW) when T_4 is OFF. This implies that an open collector (OC) output can sink current, but it cannot source current.

Q.3. Why do open collector outputs need a pull-up resistor?

Ans. In order to get an open collector output to produce a HIGH, an external resistance called a pull-up resistor should be used as shown below:

When T_4 is OFF (open) the output is nearly 5 V (HIGH) and when T_4 is ON (short) the output is nearly 0 V (or LOW). The maximum value for the pull-up resistor depends on the size of the output loads and the leakage current through T_4 when it is OFF. A good value of pull-up resistor is 10 kΩ which is neither too small to allow excessive current flow when T_4 is ON nor it is too large to cause an excessive voltage drop across itself when T_4 is OFF.

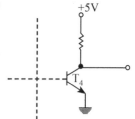

Q.4. Describe in brief, tristate switch/buffer.

Ans. At the input of a digital system, there may be more than one input signal of interest. Generally it will be necessary to connect only one signal at a time. Thus there is a requirement to connect or disconnect (switch) input signals electronically. Similarly the output of a digital system may need to be directed to more than one destination, one at a time.

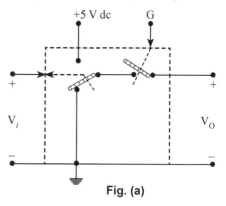

Fig. (a)

The logic circuit shown in Fig. (*a*) is a simple buffer with an additional switch controlled by an input labeled G. When G is low, this switch is open and the output is "disconnected" from the buffer. When G is high, the switch is closed and the output follows the input. That is, the circuit behaves as an ordinary buffer amplifier. In effect, the control signal G connects the buffer to the load or disconnects the buffer from the load. The truth table in Fig. (*b*) summarises circuit operation. We find that when G is high, V_O is either high or low (two states). However when G is low, the output is in effect an open circuit (a third state). Since there are three possible states for V_O, this circuit is called a tri-state buffer. Fig. (*c*) shows the symbol of tri-state buffer.

V_i	G	V_O
L	L	Open
H	L	Open
L	H	L
H	H	H

Fig. (*b*)

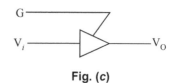

Fig. (*c*)

Q.5. Explain the operation of CMOS-NOR gate.

Ans. A circuit for a CMOS-NOR gate is shown in figure below:

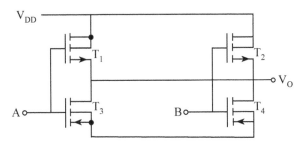

Here p-channel MOSFETs (T_1 and T_2) are connected in parallel, and the N-channel MOSFETs (T_3 and T_4) are also connected in parallel.

When both inputs A and B are low, the corresponding p-channel MOSFET *i.e.* T_1 and T_2 are turned ON and corresponding N-channel MOSFETs T_3 and T_4 are turned OFF. This causes output at Y to be equal to $+V_{DD}$ *i.e.* HIGH or 1. When A is low and B is high, then T_1 is turned ON and N-channel MOSFET T_3 is turned OFF. Also as B is high, T_2 is cut-off and N-channel T_4 is turned ON. As T_2 is not conducting and T_4 is conducting the output, the output is grounded or LOW or 0. Similarly when A is high and B is LOW, T_1 is turned OFF and T_3 is turned ON. Also B being low, T_2 is turned ON and T_4 is cut-off. As T_1 is non-conducting and T_3 is conducting, the output is grounded i.e. low or 0. However, when both A and B are HIGH, T_1 is turned OFF and T_3 is turned ON. Also T_2 is turned OFF and T_4 is turned ON which gives output Y as LOW or 0. Hence the truth table of the above circuit is as shown below

Input		Transistor state				Output
A	B	T_1	T_2	T_3	T_4	Y
0	0	ON	ON	OFF	OFF	1
0	1	ON	OFF	OFF	ON	0
1	0	OFF	ON	ON	OFF	0
1	1	OFF	OFF	ON	ON	0

The truth table is the truth table of NOR gate. This shows that the circuit in Fig. (*a*) works as a CMOS-NOR gate.

Q.6. Explain the operation of CMOS-NAND gate.

Ans. A circuit for a CMOS-NAND gate is shown below in Fig. (*a*). It makes use of p-channel MOSFETs (T_1 and T_2) connected in parallel and the N-channel MOSFETs (T_3 and T_4) connected in series. A two input CMOS-NAND gate as shown in Fig. (*a*) shows that the driver transistors are connected in series while the load transistors are connected in parallel. The inputs are applied simultaneously to a pair of transistors, one driver and other load. The output Y will be at logic 0 or LOW only when both driver transistors are ON, in which case both the load transistors will be OFF. In other words the output will be LOW (0) if A and B both are HIGH (1), the transistors T_1 and T_2 will cut-off while T_3

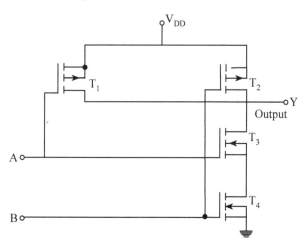

and T_4 will conduct. Now if A and B both or either of them are in state '0', then transistors T_1 or T_2 will conduct while T_3 and T_4 will be cut-off and output will be HIGH (1). Hence the truth table of the above circuit is as shown below:

Input		Transistor state				Output
A	B	T_1	T_2	T_3	T_4	Y
0	0	ON	ON	OFF	OFF	1
0	1	ON	ON	OFF	OFF	1
1	0	ON	ON	OFF	OFF	1
1	1	OFF	OFF	ON	ON	0

Fig. (*b*)

We find that the above truth table is the truth table of NAND gate. Hence the circuit in Fig. (*a*) works as a CMOS-NAND gate.

Q.7. List the differences between TTL logic family and MOS family.

Ans. Main differences between TTL logic family and MOS family are as follows:

TTL logic family	MOS family
1. Bipolar.	1. Unipolar.
2. Power supply required is 5 V.	2. Power supply can be 3 V to 15 V.
3. Propagation delay time is 15 ns.	3. Propagation delay time is higher than TTL. It varies from 25 ns to 100 ns.
4. Power dissipation is 100 mW.	4. Power dissipation around 10 mW (less than TTL)
5. Noise margin is 0.4 volts.	5. Noise margin is higher than TTL. It increases with supply voltage. If 12 V supply voltage is used, noise margin will be 5.4 V. The noise margin is roughly 0.45 V_{DD}.
6. Fan-out is 10 for TTL logic family	6. Because of very high input impedance fan-out is higher than TTL family.

LONG ANSWER TYPE QUESTIONS

Q.1. Write short note on interfacing of TTL and CMOS ICs.

Ans. Often one needs to interface between the various TTL and CMOS families. This needs to ensure that a HIGH out of a TTL gate looks like a HIGH o the input of a CMOS gate and also a HIGH out of CMOS gate looks like a HIGH to the input of a TTL gate looks like a HIGH to the input of a TTL gate. This should also hold good for a LOW logic level.

To get the optimum performance from a digital system, one has to use devices from more than one logic family, making use of superior characteristics of each family for different parts of the system. For example, TTL can be used in those portions of the system which requires high speed operation, while CMOS ICs can be used where low power dissipation is required. Hence we need to interface between TTL and CMOS ICs.

Now as the supply voltage used for all 74 series TTL ICs is 5 V, it is necessary to operate CMOS devices at +5 V to make it compatible with TTL devices.

Next let us study the problem that might arise when interfacing a standard 7400 series TTL to a 4000B series CMOS. Fig. (*a*) and (*b*) below shows the input and output specifications for TTL and CMOS correspondingly for power supply 5 V.

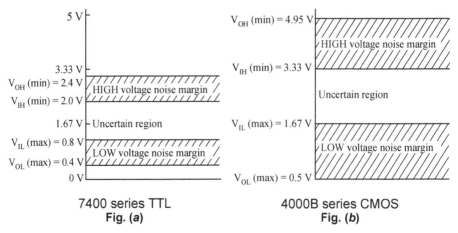

7400 series TTL
Fig. (a)

4000B series CMOS
Fig. (b)

When TTL gate is used to drive CMOS gate, there is no problem for low level output as TTL guarantees a maximum low level output of 0.4 V, this when fed to a CMOS will be low level input for CMOS as well because of the fact that any voltage upto 1.67 V is considered as LOW state by CMOS devices. But for HIGH level, the TTL output can be as small as 2.4 V as a HIGH. The CMOS interprets any voltage above 3.33 V as HIGH level input. Therefore in this case the maximum output corresponding to HIGH state of TTL of 2.4 V will be considered as LOW state input by CMOS as it follows within the uncertain region.

To solve this problem 10 kΩ resistor is connected between CMOS input to V_{CC} as shown in Fig. below:

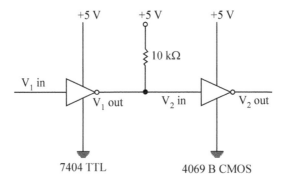

This 10 kΩ resistor is called pull-up resistor and is used to increase the output of the TTL gate closer to 5 V where it is in a HIGH output state.

Another aspect which one should look at when interfacing is the current levels of all gates that are connected which are no problem as long as TTL gate is driving CMOS gate.

In case CMOS gate is used to drive TTL gate, the voltage levels are no problem as the CMOS will output about 4.95 V for HIGH and 0.5 V for a LOW, which is easily interpreted by TTL gate.

But current levels can create problem as 4000 B CMOS has severe output current limitations. For HIGH output conditions, the 4069 B CMOS can source a maximum current of 0.51 mA which is good enough to supply the HIGH level input current to one 7404 inverter.

For LOW output conditions the 4069 B CMOS can sink only 0.51 mA which is not enough for the 7404 LOW-level input current. To overcome this problem, we use two special gates, the 4050 buffer and the 4049 inverting buffer specially designed to provide high output current to solve many interfacing problem.

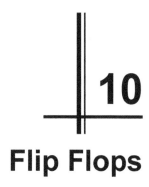

10

Flip Flops

FACTS THAT MATTER

1. **Bistable:** Any device or circuit that has two stable states is said to be bistable. For instance a toggle switch has two stable states, either up or down.

2. **Flip Flop:** Flip Flop is a bistable electronic circuit that has two stable states *i.e.* its output is either 0 or 1 (+5 V dc). The flip-flop is also regarded as a memory device since its output will remain as set until something is done to change it.

3. **Characteristic Equation:** logic expression describing a flip-flop is known as characteristic equation.

4. **Latch:** The flip-flop is often called a latch, since it will hold , or latch, in either stable state.

5. **Asynchronous:** Independent of clock pulse. The output can change without having to wait for a clock pulse.

6. **Synchronous:** Dependent on clock pulse. A clock signal must be present in order for the outputs to change states. Outputs change states in time with clock.

7. **Buffer Register:** A group of memory elements, often flip-flops, that can store a binary word.

8. **Edge-Triggering:** In this case the circuit responds only when the clock is in transition between its two voltage states.

9. **Propagation Delay:** The amount of time it takes for the output to change states after an input trigger.

10. **Set-up time:** The minimum amount of time required for data inputs to be present before the clock arrives.

11. **Hold-time:** The minimum amount of time that data must be present after the clock trigger arrives.

12. **State:** The set of memory values at any given time for a sequential logic circuit.

13. **Finite state Machine:** It is the functional description of sequential circuit.

14. **Moore Model:** Output is dependent only on current state of the circuit.

15. **Mealy Model:** Output is dependent both on current state and input to the circuit.

16. **Types of Flip-Flops:** Various types of Flip-Flops are:
    (a) RS Flip Flop, (b) D Flip Flop,
    (c) JK Flip Flop (d) T Flip-Flop

17. **Clear or Reset and Preset Function:** When power is first applied, flip-flops come up in random states. To get some computers started, an operator has to push a RESET button. This sends a CLEAR or RESET signal to all flip-flops. Also it is necessary in some digital systems to PRESET (synonymous with set) certain flip-flops.

    The PRESET and CLEAR are asynchronous inputs because they activate the flip-flop independently of the clock.

18. **Positive-edge Trigger:** When triggering occurs on the positive going edge or leading edge of the clock, it is referred to as positive-edge triggering.

19. **Negative-edge Trigger:** When triggering occurs on the negative-going edge or trailing edge of the clock, its referred to as Negative edge Trigger.

20. **SET and RESET:** When the output of a flip-flop is high or 1, we say that the flip-flop is SET. When the output of a flip-flop is low or 0, we say that the flip-flop is RESET.

21. **Excitation Table:** It is truth table written in a reverse way such that inputs are shown dependent on a particular state transition.

22. **Analysis and Synthesis of a sequential circuit:** Analysis of a sequential circuit helps to understand performance of a given circuit

in a systematic manner and through synthesis we develop circuit diagram for a specified problem.

OBJECTIVE TYPE QUESTIONS

1. A flip-flop can store _____ bit of data

2. Preset and clear inputs are synchronous inputs. (True or False)

3. There is no difference between latch and flip-flop. (True or False)

4. Flip-flop is a memory element. (True or False)

5. Race around condition is eliminated in masterslave JK flip-flop. (True or False)

6. What do the letters R and S stand for in the term "RS Latch"?

7. What is positive-edge-triggering?

8. What does an entry X mean in a flip flop truth table ?

9. Which flip-flop is easier to use, the RS or the D, as a clocked or gated latch to store data?

10. What is the primary difference between a JK and an RS flip-flop?

11. T flip-flop is known as _____ flip-flop.

12. What is characteristic equation of a flip flop?

13. How is excitation table different from flip-flop truth table?

14. What is analysis of sequential circuit?

15. Which of truth table and excitation table is useful for analysis of a sequential circuit?

16. Why flip-flop conversion is needed?

17. What is the basic difference between analysis and synthesis steps?

18. A delay flip-flop is also called _____ flip flop.

19. A negative-edge-triggered flip-flop transfers data from input to output on the High to Low transition of the clock pulse (True or false)

20. In a D flip-flop the output state Q is related with D input → Q is same as D. (True or False).

Answers

1. One   2. False.   3. False.   4. True.   5. True.

6. R stands for RESET (output Q =low). S stands for SET (Q=High).

7. When triggering occurs on the positive going edge or leading edge of the clock.

8. X means don't care –this input at this time has no effect.

9. The D flip-flop is easier to use because it requires only one input (D).

10. The JK flip-flop has an additional input condition → J = K = High. This causes the flip-flop to toggle with the clock. The R = S = High input condition is not allowed with an RS flip-flop.

11. Toggle.

12. Logic relation showing next state as a function of current state and current inputs.

13. It is truth table written in a reverse way such that inputs are shown dependent on a particular state transition.

14. Finding what a given circuit does.

15. Truth table.

16. By this one need not redesign the whole circuit if flip flop one kind is not available.

17. In analysis, problem begins with a circuit diagram and ends in state transition diagram or performance description. It uses flip flop truth table or characteristic equation in this process. In synthesis, the path is reverse and we use excitation table instead of truth table.

18. D.

19. True.

20. True.

SHORT ANSWER TYPE QUESTIONS

Q.1. What is the difference between a flip-flop and a latch?

Ans. When the flip-flop has its output set at 0 V dc, it can be regarded as storing a logic 0 and when its output is set at +5 V dc, as storing a logic 1.

Now as long as the inputs to the flip-flop are 0 or continue to be as the same input, the flip-flop will hold the last state. In this condition the flip-flop is often called a latch, because in such situation the flip-flop holds or it is latched in either of the two stable states. If the last state was set, the flipflop will remain as set and if it was reset the flip-flop will remain as reset.

When the output of a flip-flop changes because of its change in input, it is called flip-flop not latch.

Q.2. (a) Explain the working of RS flip-flop and derive its truth table using NOR gate

(b) Draw the equivalent RS flip flop using NAND gate.

Ans. The basic circuit of RS flip-flop using NOR gate is shown below:

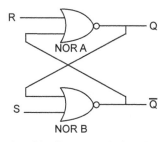

The two inputs to the flip-flop are defined as R and S. The flip-flop actually has two outputs defined as Q and $\overline{Q}$. It should be clear that regardless of the value of Q, its complement is $\overline{Q}$.

To understand the operation of the circuit, let us assume that Q = 0 and so $\overline{Q}$ = 1.

Let us now find out the input/output possibilities for the four possible inputs.

Case 1: R = 0, S = 0.

In this case since Q = 0, so both the inputs of NOR gate B is 0, so $\overline{Q}$ = 1 which is already 1. Input of NOR gate A is 0 and 1,

∴ Q = 0 which is already 0. That is when R = 0, S = 0 the flip-flop simply remains in its present state or Q remains unchanged.

Case 2: R = 0, S = 1

In this case inputs to NOR gate B is 0 and 1,

∴ $\overline{Q}$ = 0. Both the inputs to NOR gate A are now low,

∴ Q will be high. Thus a 1 at the S input is said to SET the flip-flop, and it switches to the stable state where Q = 1.

Case 3: R = 1, S = 0.

This condition forces the output of NOR gate A low that is Q = 0. Now since both the inputs to NOR gate B are now low, the output must be high or $\overline{Q}$ = 1. Thus a 1 at the R input is said to RESET the flip-flop and it switches to the stable state where Q = 0 (or $\overline{Q}$ = 1).

Case 4: R = 1, S = 1

This condition forces the output of both NOR gates to the low state since output of NOR gate is low if any or both the inputs are high. In other words, both Q = 0, and $\overline{Q}$ = 0 at the same time. But this violates the basic definition of a flip-flop that requires Q to be the complement of $\overline{Q}$. Hence this input condition is forbidden and truth table entry is?

Thus the truth table of RS flip flop is derived as shown below:

R	S	Q	Action
0	0	Last state	No change
0	1	1	SET
1	0	0	RESET
1	1	?	Forbidden

Truth Table

Logic Symbol

(b) The equivalent RS flip-flop using NAND gates is shown in fig (a) below. The NOR-gate realisation as shown above is an exact equivalent of the NAND gate realisation in Fig. (a)

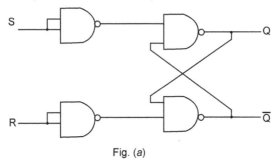

Fig. (a)

They both have the exact same symbol and truth table as given above. Both of these RS flip-flops or latches are said to be transparent, that is, any change in input information at R and S is transmitted immediately to the output at Q and $\overline{Q}$ according to the truth table.

Q.3. Describe clocked RS flip-flop using logic diagram and truth table.

Ans.

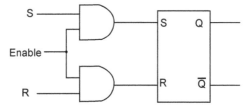

The addition of two AND gates at R and S inputs as shown in figure above will result in a flip-flop that can be enabled or disabled. When enable input is low, the AND gate outputs will both be low and so changes in neither R nor S will have any effect on the flip-flop output Q. The latch is said to be disabled in this case.

When enable input is high, information at R and S inputs will be transmitted directly to the outputs. The latch is now said to be enabled. As long as the enable is high, the output will change in response to input changes. When enable input goes low, the output will retain the information that was present on the input when high - to - low transition took place.

Thus in this way it is possible to strobe or clock the flip-flop in order to store information i.e. set it or reset it at any time and then hold the stored information for any desired period of time. This flip-flop is called gated or clocked RS flip-flop. The symbol and truth table are given below.

EN	S	R	Q_{n+1}
1	0	0	Q_n (no change)
1	0	1	0
1	1	0	1
1	1	1	? (illegal)
0	X	X	Q_n (no change)

Truth Table

Now we find that there are three inputs S, R and Enable or clock input labelled EN. Truth table output is Q_{n+1} bit simply Q. This is because we must consider two different instants in time - the time just before EN goes low Q_n and the time just after EN goes high Q_{n+1}. When EN=O, the flip-flop is disabled and R and S have no effect. Thus the truth table entry for R and S in this case is X which means don't care.

Q.4. Figure below shows the input waveforms R, S and EN applied to a clocked RS flip-flop. Explain the output waveform Q.

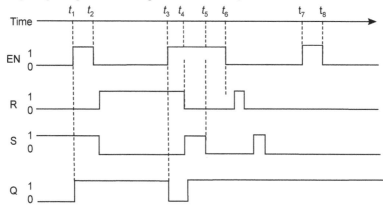

Input wave form R,S, EN applied to a clocked RS flip-flop

Ans. Initially the flip flop is reset (Q = 0) since R = 0, S = 1, EN = 0. At time t_1, EN goes high so the flip-flop is now enabled and it is immediately set (Q = 1) since R = 0 and S = 1. At time t_2, En goes low and the flip-flop is disabled and latches in the stable state Q = 1.

Between t_2 and t_3 both R and S change states, but since EN is low, the flip-flop is still disabled and Q remains at 1.

Between time t_3 and t_6 since EN is high, the flip-flop will respond to any change in R and S. Thus at t_3, Q goes low and at t_4 it goes back high according to the value of S. No change occurs at t_5, since now R and S are both low. At t_6, since EN goes low, the value Q = 1 is latched and no change occurs in Q between t_6 and t_7 even though both R and S change during this time. Between t_7 and t_8, no change in Q occurs since both R and S are low even though EN goes high at t_7.

Q.5. With the help of logic diagram, truth table and waveforms, explain the working of positive and negative edge triggered RS flip-flop.

Ans. Nearly all of the circuits in a digital system change states in synchronism with the system clock A change of state will either occur as the clock transitions from low to high called positive transition (PT) or high to low called negative transition (NT).

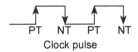

PT   NT   PT   NT
Clock pulse

A circuit that change state at positive transition of the clock is called positive-edge triggered and a circuit that changes state at negative transition of the clock is called negative-edge triggered.

Positive-edge Triggered RS flip-flop

The logic diagram, symbol, truth table and wave forms are shown below in Fig (*a*), Fig (*b*), Fig (*c*) and Fig (*d*) respectively.

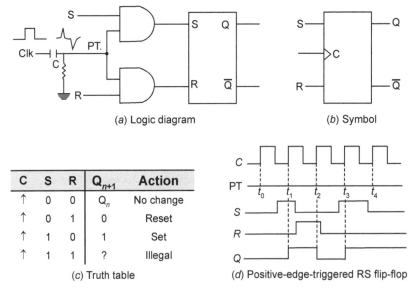

C	S	R	Q_{n+1}	Action
↑	0	0	Q_n	No change
↑	0	1	0	Reset
↑	1	0	1	Set
↑	1	1	?	Illegal

(*c*) Truth table   (*d*) Positive-edge-triggered RS flip-flop

As shown in Fig. (*a*) above, the clock is applied to a positive pulse-forming circuit. The positive spikes developed are then applied to a gated RS flip-flip. The result is a positive edge-triggered RS flip-flop with the symbol shown in

Fig (*b*). The arrow in c input indicates that Q can change state only with PTs of the clock. Each PT of the clock is a very narrow spike shown as ↑ in the truth table fig(c), that is applied to the AND gates. The AND gates are active only while the PT is high (say 25 ns) and thus Q can change state only during this short time period. In other words Q changes state in synchronism with the PTs of the clock. From the truth table we also find that input S and R affect Q only while PT occurs. From the waveforms fig(d), the above fact is clear. Also Q changes state in exact synchronism with PTs of the clock C. While drawing the wave forms, we should always take care of hold time and set time of the inputs R and S.

Negative-Edge-Triggered RS Flip-flop

The logic diagram is same as positive edge-triggered RS flip-flop. The symbol, truth table and wave forms are shown in fig (a), (b) and (c) respectively.

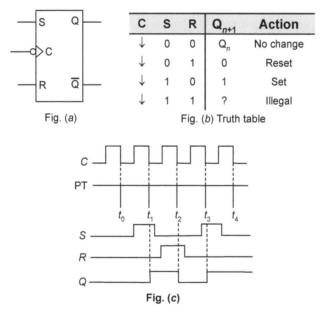

Fig. (*a*)

C	S	R	Q_{n+1}	Action
↓	0	0	Q_n	No change
↓	0	1	0	Reset
↓	1	0	1	Set
↓	1	1	?	Illegal

Fig. (*b*) Truth table

Fig. (*c*)

On the symbol, the small bubble on the clock input C means active low. This bubble, along with dynamic input indicator, means negative-edge triggering. This flip behaves exactly like positive edge-triggered R S flip-flop, except that changes in output Q are synchronized with NTs of the clock C. The truth-table and the wave forms show that Q changes state according to the R and S inputs, but only during NTs or trailing edge of the clock.

Q.6. Explain the working of clocked D flip-flop. Show its symbol and truth table.

Ans. The RS flip-flop has two data inputs R and S. To store a high bit, S should be kept high; to store a low bit, R should be kept high. Generation of two signals to drive a flip-flop is a disadvantage in many applications. Moreover the forbidden condition of both R and S high may occur inadvertently. These drawbacks are overcomed by using D flip-flop, a circuit that needs only a single data input.

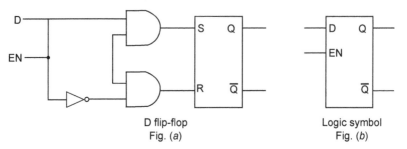

D flip-flop   Logic symbol
Fig. (a)   Fig. (b)

Fig. (a) above shows a simple way to build a D flip-flop. This flip-flop is disabled when EN is low, but is transparent when EN is high. The working of the circuit is as follows: When EN is low, both AND gates are disabled, hence change in D input will not affect Q output. On the other hand when EN is high, both AND gates are enabled. In this case Q is forced to equal the value of D. When EN again goes low, Q retains or stores the last value D. This kind of D flip-flop is often called a D latch. In general, a D flip-flop is often called a D latch . In general, a D flip-flop is a bistable circuit whose D input is transferred to the output when EN is high. Fig. (c) below shows the truth table of D flip-flop.

EN	D	Q_{n+1}
0	X	Q_n (last value)
1	0	0
1	1	1

Fig. (c) Truth table

When (EN) is low, D is a don't care (X), Q will remain latched in its last state. When EN is high, Q takes on the value of D. IF D is changing while EN is high, it is the last value of D that is stored.

Q.7. Explain the working of edge-triggered D flip-flop.

Ans. Although the D latch is used for temporary storage in electronic instruments, an even more popular kind of D flip-flop is used in digital systems, which is known as Edge-triggered D flip-flop. This kind of flip-flop samples the data bit a unique point in time.

Fig (a) below shows the circuit of the positive-edge-triggered D latch and Fig (b) shows the symbol.

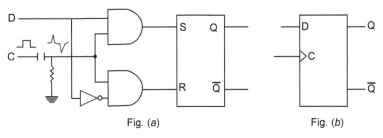

Fig. (a)   Fig. (b)

The narrow positive spike enables the AND gates for an instant. The effect is to activate the AND gates during the PT of clock, which is equivalent to sampling the value of D for an instant. At this point of time, D and its complement hit the flip-flop inputs, forcing Q to set or reset (unless Q is already equal to D). This operation is called edge-triggering because the flip-flop responds only when the clock is in transition between its two voltage states. The triggering in the Fig (a) above occurs on the positive-going edge of the clock, so it is called positive-edge triggering. The truth table summarizes the action of a positive-edge triggered D flip-flop. When the clock is low, D is don't care and Q is latched in its last state. On the other hand, on the leading edge of the clock (PT) designated by the up-arrow, the data bit is loaded into the flip-flop and Q takes on the value of D.

C	D	Q_{n+1}
0	X	Q_n (last state)
↑	0	0
↑	1	1

Truth table

Q.8. Typical waveforms for setting a 1 in a positive edge-triggered D flip-flop are shown in Figure below. Discuss the timing.

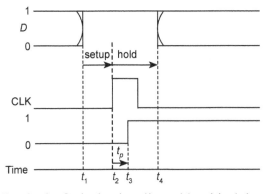

Ans. The lower line in the fig is the time line with critical times marked on it. Prior to t_1, the data can be a 1 or a 0 or can be changing. This is shown by drawing lines for both high and low levels of D. From time t_1, to t_2, the data D must be held steady at 1 (since we have to store 1). This is the setup time t set up. Data is shifted into the flip-flop at time t_2 but does not appear at Q until time t_3. The time t_2 to t_3 is the propagation delay time t_p. In order to guarantee proper operation, the data line must be held steady from t_2 until t_4. This is the hold time t_{hold}. After t_4, D is free to change states, shown by the double lines.

Q.9. Explain the working of positive-edge-triggered JK flip-flop.

Ans. We have seen that setting R = S = 1 with an edge-triggered RS flip-flop forces both Q and $\overline{Q}$ to the same logic level which is illegal and not possible. The JK flip-flop accounts for this illegal input and is therefore a more versatile circuit.

One way to build a positive-edge-triggered JK flip-flop is shown below:

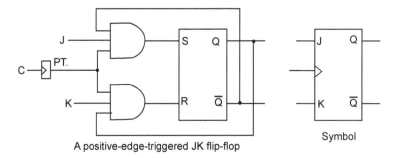

A positive-edge-triggered JK flip-flop

The pulse-forming box changes the clock into a series of positive pulses and thus this circuit will be sensitive to PTs of the clock. Here the Q output is connected back to the input of the lower AND gate and $\overline{Q}$ is connected back to the input of the upper AND gate.

This cross coupling from outputs to inputs changes the RS flip-flop into a JK flip-flop. The S input is now labeled J and R input is labeled K. Here is how the circuit works:

1. When J and K are both low, both AND gates are disabled. Hence clock pulses have no effect and Q retains its last value.
2. When J=low and K = high, the upper AND gate is disabled. When Q is high, the lower AND gate passes a RESET pulse as soon as the next positive clock edge arrives. This forces Q to become low. Therefore J = 0 and K = 1 means that the next PT of the clock resets the flip-flop (unless Q is already reset).
3. When J is high and K is low, the lower AND gate is disabled. If Q is low, then $\overline{Q}$ is high, therefore the upper AND gate passes a SET pulse on the next positive clock edge. This forces Q to become high. So J = 1 and K = 0 means that the next PT of the clock sets the flip-flop (unless Q is already set).

4. When J and K are both high, it is possible to set or reset the flip-flop. If Q is high, the lower gate posses a RESET pulse on the next PT of the clock. On the other hand if Q is low, the upper gate passes a SET pulse on the next PT of the clock. That means either way, Q changes to the complement of the last state. So J = K = 1 means flip-flop will Toggle (switch to the opposite state) on the next PT of the clock.

The Truth table is shown below.

C	J	K	Q_{n+1}	Action
↑	0	0	Q_n	No change
↑	0	1	0	Reset
↑	1	0	1	Set
↑	1	1	$\overline{Q}_n$	Toggle

We can notice here that propagation delay prevents the JK flip-flop from racing (toggling more than once during a positive clock edge). The outputs change after the PT of the clock. By then, the new Q and $\overline{Q}$ values are too late to coincide with the PTs driving the AND gates. For e.g. if t_p = 20ns, the outputs change approximately 20 *ns* after the leading edge of the clock. If PTs are narrower than 20 *ns*, the returning of Q and $\overline{Q}$ is too late to cause false triggering.

Q.10. Explain the working of JK Master slave flip-flop. Also how the racing problem is avoided in this Flip Flop?

Ans. JK Master-slave flip-flop requires two JK flip-flops—one is called master and the other is called slave. The Master is positive edge-triggered and the slave is negative-edge-triggered. Master responds to J and K first and then slave copies the same. Figure below shows the JK master slave arrangement:

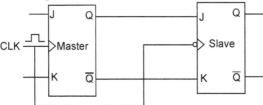

This is how the circuit works:

1. If J = 1 and K = 0, the master sets as soon as positive edge of the clock signal occurs forcing the output Q of master to be high. The high output (Q = 1) of the master drives the J input of the slave and when a negative edge of the clock signal occurs, the slave sets or copies the master.

2. If J = 0 and K = 1, the master resets on the positive edge of the clock signal which forces $\overline{Q}$ of master to be 1 and Q = 0. The high $\overline{Q}$ output of master drives the K input of the slave and at the occurrence of negative edge of the clock signal, the slave repeats the same action of the master *i.e.* slave is reset.

3. If J = 1 and K = 1 i.e. both inputs are high for the master, it toggles at the positive edge of the clock signal and slave toggles at the negative edge of the clock signal copying the master again.

4. If J = K = 0 the flip-flop is disabled and Q remains unchanged.

The Truth-Table is shown below:

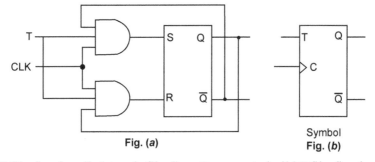

C	J	K	Q_{n+1}	Action
⊓	0	0	Q_n	No change
⊓	0	1	0	Reset
⊓	1	0	1	Set
⊓	1	1	$\overline{Q}_n$	Toggle

Symbol

From the above description and truth-table we find that in JK master-slave flip-flop the problem of racing (toggling more than once during a positive clock edge) does not arise at all. Here the racing problem is avoided without using RC network, since here the master is positive edge triggered and the slave is negative-edge triggered. In the same clock pulse, the master responds first at the positive edge and slave copies the same at the negative edge. Then only the second clock signal occurs. So, there is no racing problem using JK master/slave flip-flop.

Q.11. Explain the working of T flip-flop and derive its truth table.

Fig. (a)   Symbol
                                                  Fig. (b)

T flip-flop is called toggle flip-flop. One way to build T flip-flop is shown in the fig (a) above and the symbol is shown in fig (b). In such a flop-flop, narrow triggers drive the T input. Each time one trigger appears, the output of the flip-flop changes states or the flip-flop toggles. For instance, suppose Q equals 0 just before the trigger, then as soon as the trigger arrives, the upper AND gate is enabled and results in a High S input. This sets the Q output to 1. When the next trigger arrives the lower AND gate is now enabled and so it results in R input to be high which forces the flip-flop to reset or $\overline{Q}$ output 1 and Q = 0. As long as T input is high or 1, on arrival of clock pulse, the output is changed to its complement. The truth table is shown below:

C	T	Q	Q_{n+1}
↑	0	0	0
↑	0	1	1
↑	1	0	1
↑	1	1	0

In Fig (*a*) shown above actually T flip-flop is built by tieing together J and K input of the JK flip-flop. It is a single input version of the JK flip flop. Here J = K = T.

Q.12. Convert RS flip-flop to clocked D flip-flop, T flip-flop and JK flip-flop.

Ans. It is usually possible to convert one type of flip-flop to another by adding external gates if required.

Conversion of RS flip-flop to clocked D flip-flop

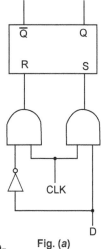

Fig. (*a*) shows the desired behavior of clocked D flip-flop after the RS flip-flop has been connected as shown. The flip-flop is set to 1 if D = 1 and a clock pulse occurs. Therefore S = D (CLK), Similarly the flip-flop must be set to 0 if D = 0 and a clock pulse occurs. Therefore R = $\overline{D}$ (clk).

Thus RS flip-flop is converted to clocked D flip-flop.

Conversion of RS flip-flop to T flip-flop

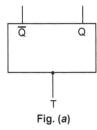

Fig. (*a*)

The trigger or T flip-flop as shown in the symbol in Fig. (*a*) has a single input. Applying a pulse to this input causes it to change state. For example if Q = 1 and the T input is pulsed, Q changes to 0. If T is pulsed again, Q changes back to 1. R S flip-flop can be converted to T flip flop by adding two AND gates in the following manner shown in Fig. (*b*).

Fig. (*a*)

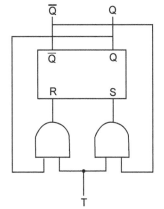

Fig. (*b*)

Here we find that R = QT and S = QT and so this RS flip-flop behaves like T flip-flop. Thus RS flip-flop is converted to T flip-flop.

Conversion of RS flip-flop to JK flip-flop

The J-K flip-flop combines the features of the RS and T flip-flop. That is if J = 1, the flip-flop output is set to Q = 1. If K = 1, the flip-flop output is reset to Q = 0. If J = K = 1 the flip-flop changes state just like T flip-flop. For example if Q = 0 and we momentarily apply a 1 input to both J and K, the flip-flop state will change to Q = 1.

Fig. (a) below shows the symbol of JK flip-flop and Fig (*b*) shows how RS flip-flop can be converted to JK flip-flop with added gates.

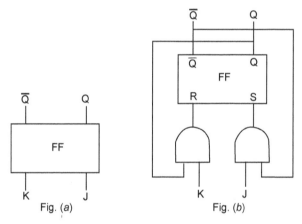

Fig. (*a*)   Fig. (*b*)

In Fig (b) we find that if Q = 0, an input of J = 1, K = 0 will set the flip-flop to Q = 1. If Q = 1, and input of K = 1, J = 0 will reset the flip flop to Q = 0. If J = K = 1, then the flip-flop acts like a T flip flop and a state change occurs.

Thus RS flip-flop is converted to J-K flip-flop.

LONG ANSWER TYPE QUESTIONS

Q.1. Derive the characteristic (next state) equation for the SR flip flop, T flip-flop, JK flip flop and D flip-flop.

Ans. The characteristic equation or next state equation for the flip flop can be derived as follows:

First make a truth-table which gives the next state (Q^+) as a function of the present state (Q) and the inputs. Any illegal input combinations should be treated as don't cares. Then plot the K-map for Q^+ and find the characteristic equation from the map.

For S-R Flip-flop

The behaviour of the S-R flip-flop is summerised by the truth table shown below:

S(*t*)	R(*t*)	Q(*t*)	Q(*t* + ∈)
0	0	0	0
0	0	1	1
0	1	0	0
0	1	1	0
1	0	0	1
1	0	1	1
1	1	0	− ⎤ Inputs not
1	1	1	− ⎦ allowed

In the above table Q(t) represents the "present state" of the flip flop, ∈ is the time required for a change of state to occur, and Q (*t* + ∈) is the "next state" of the flip-flop. Now we shall draw the Karnaugh map for Q(*t* + ∈).

$$Q(t + \in) = S(t) + R(t)Q(t)$$

Normally we will write this equation without including time explicitly using Q to represent present state of the flip flop and Q+ to represent the next state .

$$Q^+ = S + R'Q \quad (SR = 0) \tag{1}$$

In words, this equation tells us that the next state of the flip flop will be 1 either if it is set to 1 with an S input or if the present state is 1 and it is not reset. The condition SR = 0 implies that S and R cannot both be 1 at the same time.

An equation which expresses the next state of a flip-flop in terms of its present state and inputs is known as next state equation or characteristic equation.

Thus equation (1) is the characteristic equation of S-R flip-flop.

For T flip-flop

The behaviour of the T flip-flop is summarised by the truth table shown is figure below:

T	Q	Q⁺
0	0	0
0	1	1
1	0	1
1	1	0

	00	01
0	0	①
1	①	0

K-map

$$Q(t + \epsilon) = T'(t)Q(t) + T(t)Q'(t)$$

Or

$$Q^+ = T'Q + TQ' = T \oplus Q$$

From the truth table it is clear that if T remains 1, then the flip-flop will change state and it will continue to oscillate until T becomes 0. So, we find the characteristic equation for T flip-flop is

$$Q^+ = T'Q + TQ'$$

This characteristic equation states that the next state of the flip-flop (Q^+) will be 1 if and only if the present state (Q) is 1 and no T pulse occurs or the present state is 0 and a T pulse occurs.

For JK flip flop

The behavior of the JK flip-flop is summarised by the next state table shown below:

J	K	Q	Q⁺
0	0	0	0
0	0	1	1
0	1	0	0
0	1	1	0
1	0	0	1
1	0	1	1
1	1	0	1
1	1	1	0

From the above table we find that JK flip flop combines the features of the S-R and T flip-flop. A 1 input applied to J or K alone acts exactly like an S or R input respectively. If J = 1, the flip-flop output is set to Q = 1 and if K = 1, the flip flop output is reset to Q = 0. If J = K = 1, the flip-flop changes state just like T flip-flop i.e. it toggles.

Now we shall draw the Karnaugh Map for Q^+.

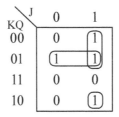

$$Q^+ = JQ' + JK' + K'Q$$
$$= JQ' + K'Q \text{ (Using consensus theorem JK' is deleted)}$$

Thus the characteristic equation for JK flip-flop is

$$Q^+ = JQ' + K'Q$$

For D flip flop

The state of D flip flop after the clock pulse (Q^+) is equal to the input (D) before the clock pulse. For instance, if D = 1 before the clock pulse, $Q^+ = 1$ after the clock pulse regardless of the previous value of Q. The next state table is shown below.

D	Q	Q^+
0	0	0
0	1	0
1	0	1
1	1	1

Therefore the characteristic equation is $Q^+ = D$.

Q.2. How you can represent SR, D, JK and T flip-flop as finite state machine through their state transition diagrams?

Ans. In a sequential logic circuit the value of all the memory elements at a given time define the state of that circuit at that time.

Finite State Machine (FSM) concept offers a better alternative to truth table in understanding progress of sequential logic with time. For a complex circuit a truth table is difficult to read as its size becomes too large. In FSM, functional behaviour of the circuit is explained using finite number of states. State transition diagram is a very convenient tool to describe a FSM.

Now let us write the truth table of each Flip flop and represent as finite state machine through their state transition diagrams

(a) SR flip flop

S	R	Q	Q⁺
0	0	0	0
0	0	1	1
0	1	0	0
0	1	1	0
1	0	0	1
1	0	1	1
1	1	0	− } Inputs not
1	1	1	− } allowed

Truth table of SR SR flip-flop

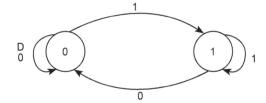

State transition diagram of SR flip-flop

Now let us describe how state transition diagram for SR flip-flop is developed from its truth table. Each flip-flop can be at either 0 or 1 state defined by its stored value at any given time. Application of input may change the stored value *i.e.* state of the flip-flop. This is shown by directional arrow and the corresponding input is written alongside. If SR flip-flop stores 0, then for SR = 00 or 01, the stored value does not change. For SR=10, flip-flop output changes to 1. When SR flip-flop stores 1, application of SR = 00 or 10 does not change its value. SR = 01 will change the output to 0. Note that SR =11 is not allowed in SR flip-flop and it is not appearing in the state transition table.

The state transition diagrams are developed in a similar way for D, JK and T flip-flop and are shown below:

(b) D flip-flop

D	Q	Q⁺
0	0	0
0	1	0
1	0	1
1	1	1

Truth table

State transition diagram of D flip-flop

(c) T Flip Flop

T	Q	Q⁺
0	0	0
0	1	1
1	0	1
1	1	0

Truth table

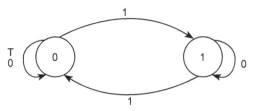

State transition diagram for T flip-flop

(d) For JK flip-flop

J	K	Q	Q⁺
0	0	0	0
0	0	1	1
0	1	0	0
0	1	1	0
1	0	0	1
1	0	1	1
1	1	0	1
1	1	1	0

Truth table

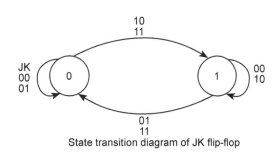

State transition diagram of JK flip-flop

The development of all the state transition diagrams for the flip-flop from its truth table may be described in the similar way as described for the SR flip-flop.

Q.3. Derive the Excitation Table for SR, JK, D and T flip-flop.

Ans. In analysis problem truth table is important. In synthesis or design problem Excitation Table is important. Excitation table of a flip-flop is looking at its truth table in a reverse way. Here flip-flop input is presented as a dependent function of transition $Q_n \to Q_{n+1}$ and comes later in the table. This can be derived from flip-flop truth table or characteristic equation but it can more easily be derived from its state transition diagram.

Now let us derive the Excitation table for each flip-flop from its state transition diagram.

SR Flip-Flop

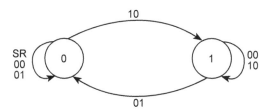

State transition diagram for SR flip-flop

Here we find if present state is 0, application of SR = 0X does not alter its value where X denotes don't care condition in R input. State 0 to 1 transition occurs when SR = 10 is present at the input side while 1 to 0 transition occurs if SR = 01. Present state 1 is maintained if SR = X0.

Hence the excitation table for SR flip-flop is

$Q_n \rightarrow Q_{n+1}$		S	R
0	0	0	X
0	1	1	0
1	0	0	1
1	1	X	0

Excitation table for other flip-flop are obtained in a similar way.

JK Flip Flop

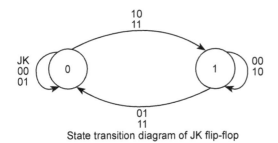

$Q_n \rightarrow Q_{n+1}$		J	K
0	0	0	X
0	1	1	X
1	0	X	1
1	1	X	0

State transition diagram of JK flip-flop   Excitation table for JK flip-flop

D Flip Flop

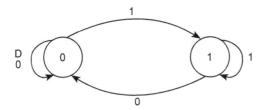

$Q_n \rightarrow Q_{n+1}$		D
0	0	0
0	1	1
1	0	0
1	1	1

State transition diagram for D flip-flop   Excitation table for D flip-flop

T Flip Flop

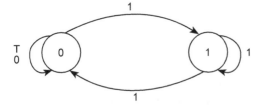

$Q_n \rightarrow Q_{n+1}$		T
0	0	0
0	1	1
1	0	1
1	1	0

State transition diagram for T flip-flop   Excitation table for T flip-flop

11

Shift Registers

FACTS THAT MATTER

1. **Register:** A register is a group of flip-flops that can be used to store a binary number. each flip-flop can store 1 bit. Therefore to store an 8-bit binary number the register must consist of eight flip-flops.

2. **Shift Register:** A group of lip-flops connected in such a way that a binary number can be shifted into or out of the flip-flops is known as shift Register.

3. **Serial Shift:** Data bits are shifted one after the other in a serial fashion with one bit shifted at each clock transition. Therefore n clock transitions are needed to shift an n-bit binary number.

4. **Parallel Shift:** Data bits are shifted simultaneously with a single clock transition.

5. **Register Capacity:** Determined by the number of flip-flops in the register.

6. **Ring Counter:** A basic shift register with direct feedback such that the contents of the register simply circulate around the register when the clock is running.

7. **Sequence Detector:** Detects a binary word from input data stream.

8. **Sequence Generator:** Generates a binary data sequence.

9. **Serial Adder:** Converts parallel data to serial and use adder block sequentially.

10. **UART :** Universal Asynchronous Receiver Transmitter UART is a clip used to exchange data in a microprocessor system. The UART is constructed using registers and some control logic.

11. **Types of shift Register:** There are four basic types of shift Register:-
    (*i*) Serial-in-Serial-out (SISO) → e.g. 74LS91, 8 bits
    (*ii*) Serial-in-parallel-out (SIPO) → e.g. 74164, 8 bits
    (*iii*) Parallel-in-serial-out (PISO) → e.g. 74165, 8 bits.
    (*iv*) Parallel in-parallel out (PIPO) → e.g.74198, 8 bits.

12. **Universal Shift Register:** A universal shift Register can perform all the four basic operations (SISO, SIPO, PISO and PIPO) and is also bidirectional in nature i.e. shift right or shift left is also possible *e.g.* 74194, 4 bit and 7495 A, 4 bit are the examples of universal shift Register.

13. **Switched tail counter or Johnson counter:** Shift register with complemented output of the last flip-flop fed to the first flip-flop input is known as switched tail counter or twisted tail counter or Johnson counter.

14. **Power-on-reset circuit:** A power-on reset circuit is used to preset flip-flops to any desired states.

OBJECTIVE TYPE QUESTIONS

1. A 4-bit shift register is to be made using D flip-flops. How many flip-flops will be required?

2. SISO, SIPO, PISO and PIPO stands for _____, _____, _____ and _____.

3. Which of the following forms can be used for transmitting data ? a) Serial form, b) parallel form, c) parallel or serial form.

4. For transmitting whole data words at the some time we will make use of which mode?

5. For shifting a 16 bit binary number into a 16 flip-flop serial shift register how many clock pulses will be required?

6. A shift register is a combinational logic circuit or sequential logic circuit?

7. How many clock pulses are required to parallel load four bits in a 4 flip-flop shift register?

8. What is a ring counter?

9. How does a serial adder work?

10. What is universal shift Register?

Answer

1. Four.
2. Serial-in-serial out, serial-in-parallel-out, parallel-in-serial-out and parallel-in-parallel-out.
3. (c) Parallel or serial form.
4. Parallel data transmission mode.
5. 16.
6. Sequential logic circuit.
7. one.
8. A ring counter is a basic shift register with direct feedback such that the contents of the register simply circulate around the register when the clock is running.
9. Serial adder converts parallel data to serial and use adder block sequentially.
10. A universal shift register can perform all the four basic operations (SISO, SIPO, PISO and PIPO) and is also birdirectional in nature.

LONG ANSWER TYPE QUESTIONS

Q.1. Explain the concept of serial-in-serial-out with the help of an example.

Ans. Serial-in-serial-out refers to data serially entering the shift register and serially exiting out of the shift register, one bit at a time. Let us consider four D flip-flops connected as shown forming 4 bit shift register in Fig. (*a*) and let a binary waveform as shown along D of Fig. (*b*) be fed to serial data input of the shift register.

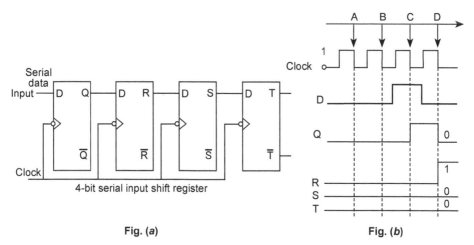

Fig. (*a*)   Fig. (*b*)

As shown in Fig. (*a*) above, a common clock provides trigger at its negative edge to all the flip-flops. As output of one D flip-flop is connected to input of the next, at every clock trigger data stored in one flip-flop is transferred to the next. For this circuit, transfer takes place like this → Q → R, R → S, S → T and serial data input is transferred to Q.

Now assume all the flip-flops are initially cleared. Now let a binary waveform as shown along D of Fig (*b*) be fed to serial data input of the shift register. Corresponding Q, R, S, T are also shown in the figure.

At clock edge A, flip-flop Q has input 0 from serial input data in D, flip-flop R has input 0 from output of Q, flip-flop S has input 0 from output of R and flip-flop T has input 0 from output of S. When clock triggers, we have QRST = 0000. Thus at clock trigger values at DQRS is transferred to QRST.

At clock edge B, serial data in = 0 *i.e.* DQRS = 0000. So after NT at B, QRST = 0000. Serial data becomes 1 in next clock cycle.

At clock edge C, serial data in = 1 *i.e.* DQRS = 1000 and after NT, QRST = 1000.

At clock edge D, serial data in = 0 *i.e.* DQRS= 0100 and after NT, QRST = 0100.

Thus the binary number 0100 is entered serially in the shift-register shown in Fig. (*a*).

Now for serial-out→let us take the same shift register as shown in Fig. (*a*) and let us assume that it has the 4-bit numbers QRST = 1010 stored in it. If clock signal is applied, the waveforms shown in Fig. (*c*) below will be generated. Here's what happens.

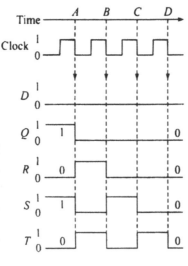

Fig. (c)

Before time A: The register stores the number QRST = 1010. The LSB (O) appears at T.

At time A: The entire number is shifted one flip-flop to the right. A 0 is shifted into Q, 1 of Q is shifted to R, 0 of R is shifted to S, 1 of S is shifted to T and 0 of T i.e. the LSB is shifted out the right end and lost. The register holds the bits QRST = 0101, and the second LSB (1) appears at T.

At time B: The bits are all shifted one flip-flop to the right, a 0 shifts into Q and the third LSB (O) appears at T. The register holds QRST = 0010.

At time C: The bits are all shifted one flip-flop to the right. A 0 is shifted into Q and the MSB (1) appears at T. The register now holds QRST = 0001.

At time D: The MSB is shifted out the right end and lost, a 0 is shifted into Q and the register holds QRST = 0000.

Thus we find that the number stored in the register is serially shifted out, 1 bit at a time beginning with the LSB, over a time period of four clock cycles. Also since the serial data input is 0 as shown by D waveform, at every clock going negative, a 0 is shifted into Q (left most flip-flop). Thus, this register is called serial in-serial out shift register.

Q.2. The content of 4 bit register is initially 1101. The register is shifted five times to the right with serial I/P being 10110. Give the table of the contents of register after each shift.

Ans. Let us take D flip-flops to construct the 4 bit register as shown in Fig. (*a*) and clock pulse in Fig. (*b*).

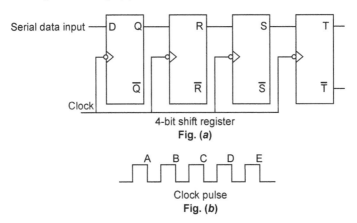

4-bit shift register
Fig. (a)

Clock pulse
Fig. (b)

Initially it is given that the content of 4-bit register is 1101 or QRST = 1101. The serial I/P to be shifted in is given as 10110.

Now as we can see from Fig. (*a*), the flip-flop Q,R,S,T are Negative Triggered. So, with every clock going negative, the content will shift 1 bit to the right. In other words shifting will occur at A,B, C, D and E of the clock pulse as shown in Fig. (*b*).

Initially QRST = 1101. Just after NT at A, LSB (0) from serial I/P 10110 will shift to Q, 1 of Q will shift to R, 1 of R will shift to S, 0 of S will shift to T and I of T will shift out and lost. Hence AT clock edge A. QRST = 0110.

In a similar way

At clock edge B, serial Data input is 1 *i.e.* DQRS = 1011. So after NT at B, QRST = 1011

At clock edge C, serial data input is 1 *i.e.* DQRS = 1101, so just after NT at C, QRST = 1101

At clock edge D, serial data input is 0 *i.e.* DQRS = 0110, so just after NT at D, QRST = 0110

At clock edge E, serial data input is 1 *i.e.* DQRS = 1011, so just after NT at E, QRST = 1011.

Thus after the register is shifted five times to the right, according to the given serial data input, the content of the 4 bit shift register will be 1011.

The table of the contents of register after each shift is shown below:-

Clock	Serial in put	Q	R	S	T
0	0	1	1	0	1
1	1	0	1	1	0
2	1	1	0	1	1
3	0	1	1	0	1
4	1	0	1	1	0
5		1	0	1	1

Data transfer through serial input in a shift register

Q.3. With the help of pin diagram and logic diagram explain the working of IC 74LS91 shift register.

Ans. The pin out and loci diagram for a 74LS91 shift register are shown in Fig (*a*) and Fig. (*b*) below. IC 74LS91 is an 8 bit TTL MSI chip. This is serial

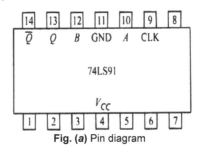

Fig. (a) Pin diagram

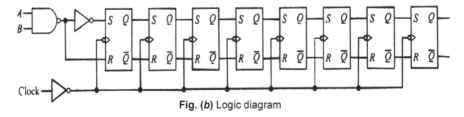

Fig. (b) Logic diagram

in serial out shift register. There are eight RS flip-flops connected to provide a serial input as well as serial output. The clock input at each flip-flop is negative-edge-trigger-sensitive. However, since the applied clock signal is passed through an inverter, data will be shifted on the positive edges of the input clock pulses.

The inverter connected between R and S on the first flip-flop means that this circuit functions as a D-type flip-flop. So, the input to the register is a single line on which the data to be shifted into the register appears serially. The data input is applied at either A (pin 10) or B (pin 12). We see that the data level at A (or B) is complemented by the NAND gate and then applied to R input of the first flip-flop. The same data level is complemented by the NAND gate and again complemented by the inverter before it appears at the S input. So, a 1 at input A will set the first flip-flop or this 1 is shifted into the first flip-flop on a positive clock transition.

The NAND gate with inputs A and B simply provides a gating function for the input data stream if desired. If gating is not desired, simply pins 10 and 12 may be connected together and the input data stream may be applied to this connection.

Thus at every positive-edge-trigger of the clock, 1 bit will shift in the first flip-flop and the whole contents of the 8 bit shift register will shift one flip-flop to the right. If there are 8 bits input to be shifted in, it will require 8 clock pulses. Similarly to take the contents out of this 8 bit register again 8 clock pulses will be required, since with every positive edge trigger one bit will shift out.

Thus this IC 74LS91 acts as serial in serial out shift register. Also the largest number that can be stored in 74LS91 register is $11111111_2 = 255_{10}$.

Q.4. With the help of pinout, logic diagram and waveforms explain the working of 74164 shift register.

Ans. The 74164 is an 8 bit serial-input-parallel-output shift register.

In order to shift the data out in parallel, it is simply necessary to have all the data bits available as outputs at the same time which is possible by connecting the output of each flip-flop to an output pin.

The pinout and logic diagram for 74164 shift register is given below:

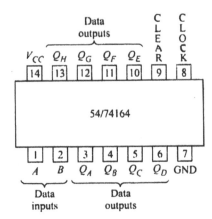

Fig. (a) DIP pinout

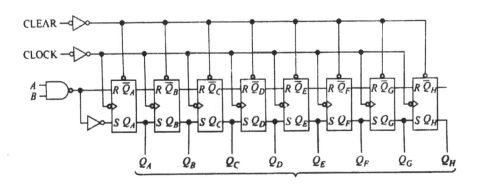

Fig. (b) Logic diagram

It is constructed by using RS flip-flops having clock inputs that are sensitive to NTS. If we observe carefully we find that the logic diagram of 74164 is exactly like that of 74LS91 with two exceptions:

1. Then true side of each flip-flop is available as an output-thus we can have all the 8 bits of any number stored in the register simultaneously which is required for parallel output,

2. Each flip-flop has an a synchronious clear input. Thus a low level at the clear input to the chip (pin 9) is applied to reset or clear every flip-flop irrespective of the clock. As long as the clear input to the chip is held low, the flip-flop outputs will all remain low or the register will contain all zeros.

The setup time is 30 ns minimum and hold time is 0.0ns for 74164, hence data at the serial input should be present at least 30 *ns* before the PT of the clock and must be stable until the clock transition is complete. Thus with every PT of clock, 1 bit from data input will be shifted in the first flip-flop. Now let us examine the gated serial inputs A and B. If the serial data is connected to A, then B can be used as a control line.

B is held high: The NAND gate is enabled and the serial input data passes through the NAND gate inverted which is again inverted by the inverter before reaching S input of the first flip-flop. Thus input data is shifted into the register with clock going positive.

B is held low: The NAND gate output is forced high, the input data stream is inhibited and the next positive clock transition will shift a 0 into the first flip-flop. Thus after eight clock pulses, the register will contain all zeros.

Now let us draw the waveforms and see the response of 74164 if say 00101100 serial data input at A has to be shifted in the register. The waveforms are shown below.

According to the waveforms, the first clear pulse occurs at time A which resets all the flip-flop to 0.

The clock begins at time B, but first PT does nothing since the control signal is low.

At time C, the control signal goes high and remains high. So first data bit i.e. LSB (O) is shifted into the register at the next PT *i.e.* at time D.

The next 7 data bits are shifted in, in order at times E, F, G, H, I, J and K. The clock remain high after time K, and the 8 bit number 00101100 is now stored in the register and is available on the eight output lines.

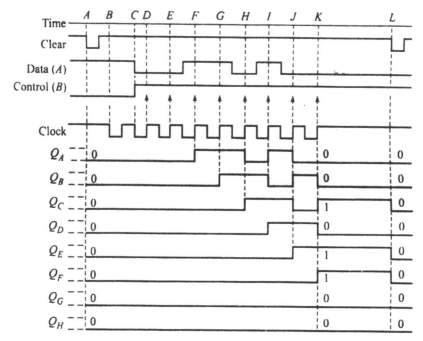

Waveforms

The clock must be stopped after its positive transition at time K, else shifting will continue and data bits will be lost.

Finally, another clear pulse occurs at time L when the flip-flops are all reset to 0 The register can otherwise be cleared by holding the control line at B low and allowing the clock to run for eight PTs. This will shift eight 0s into the register.

Q.5. IC 74166 can function as parallel in serial-out as well as serial-in-serial-out shift register. Explain how?

Ans. The pin out and logic diagram of IC 74166 is shown below:

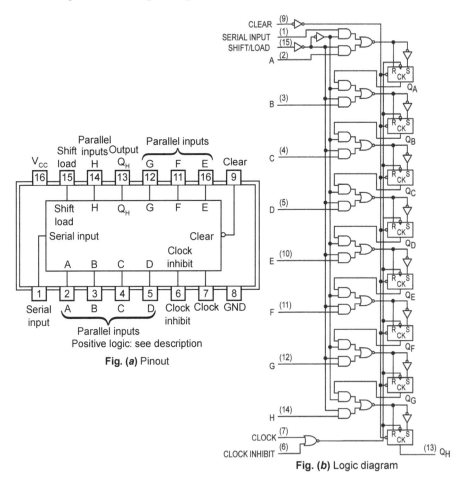

Fig. (a) Pinout

Fig. (b) Logic diagram

The 74166 is an eight bit shift register and if we observe properly, the same circuit is repeated eight times. This 8-bit shift register is capable of either parallel or serial data entry and serial data output. From the logic diagram we find that there are 8 RS flip flops each with similar logic circuitry attached to it. Let us first analyse one of these circuits as shown below:

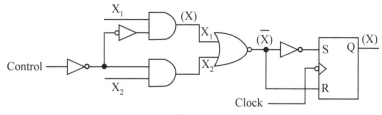

Fig. (c)

If the control line is high the upper AND gate is enabled and the lower AND gate is disabled. Thus X_1 will appear at the upper leg of the NOR gate while the lower leg of the NOR gate will be at ground level or zero. Therefore X_1 will appear at the output of NOR gate which is again inverted by the inverter before reaching S input of the RS flip-flop or data at X_1 will be shifted into the flip-flop.

On the other hand if control line is low, the upper AND gate is disabled while the lower AND gate is enabled. So now X_2 appears at the lower leg of the NOR gate while the upper leg of the NOR gate is at ground level or zero. Now X_2 will be the output of NOR gate which is again inverted by the inverter before reaching S input of the RS flip flop or data at X_2 will be shifted into the flip-flop.

To summerize: *Control is high:* Data bit at X_1 will be shifted into the flip-flop at the next clock transition.

Control is Low: Data bit at X_2 will be shifted into the flip-flop at the next clock transition.

Now if we observe properly, we find that exactly eight of the circuits given in Fig.(c) above are connected together to form 74166 shift register. They are connected to allow two different operations (i) the parallel entry of data and (ii) serial entry of data.

Referring to the logic diagram of 74166, if the control line labeled shift/load is held low, the lower AND gate for each flip-flop is enabled and the 8 bit number labeled A, B, C, D, E, F, G and H will be loaded into the flip-flops with a single clock transition. This is the parallel data entry.

Holding control or shift/load high will enable the upper AND gate for each flip-flop. The input of this upper AND gate receives its data from the prior flip-flop in the register, so each clock transition will shift a data bit from one flip-flop into the following flip-flop, proceeding from Q_A toward Q_H. Thus data will be entered and shifted serially through the register. In the first flip-flop in the register, the upper AND gate input is labeled serial input. thus data can also be entered into the register serially.

Thus **Shift/Load is low:** A single clock transition loads 8 bits of data (ABCDEFGH) into the register in parallel.

Shift/Load is High: Clock transitions will shift data through the register serially, with entering data applied at the serial input.

We also find that the clock is applied through a two input NOR gate. When clock inhibit is held low, the clock signal passes through the NOR gate inverted. Since the register flip-flops respond to NTs, data will shift into the register on the PTs of the clock.

When clock inhibit is high, the NOR gate output is held low and the clock is prevented from reaching the flip-flops. In this mode, the register can be made to stop and hold its contents.

A low level at the clear input can be applied at any time without regard to the clock and it will immediately reset all the flip-flops to 0. When not in use, it should always be held high.

Q.6. With the help of logic diagram of IC 74174 explain the working.

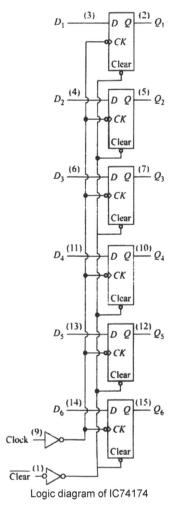

Ans. IC 74174 is a parallel in-parallel out shift register. The logic diagram of IC 74174 is shown below. It is simply a parallel arrangement of six D-type flip-flops. The six data bits are D_1 through D_6. Each flip-flop is negative edge-triggered, thus according to the logic diagram a positive transition will shift data bits D_1 through D_6 into the register in parallel. The stored data is immediately available in parallel, at the outputs Q_1, through Q_6. This type of register is used to store data and is called a data register or data latch. In this register it is not possible to shift stored data either to the right or to the left. A low level at the clear input will immediately reset all the flip-flops or all flip-flops will have 0.

The clear input is asynchronous *i.e.* independent of clock and can be used to reset the flip flops as and when required. The set up time for IC 74174 is 20 ns and a hold time is 5 *ns*.

Q.7. Explain the working of universal shift register IC 7495 A.

Ans. The four possible operations possible for basic types of shift registers are serial-in-serial-out (SISO), serial-in-parallel-out

Logic diagram of IC74174

(SIPO), Parallel in-serial out (PISO) and Parallel-in-Parallel out (PIPO). Serial in or serial out again can be shift right or shift left. A universal shift register can perform all the four operations and is also bidirectional. 7495 A is such kind of universal shift register. The pin diagram and logic diagram are shown below in Fig. (*a*) and Fig. (*b*)

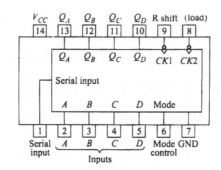

Fig. (a) Pinout

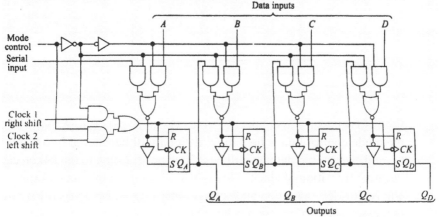

Fig. (b) Logic diagram

7495 A is a 4-bit universal shift register. The parallel data outputs are simply the Q sides of each of the four RS flip-flop in the register. In fact Q_D can also be used as a serial output when data is shifted from left to right through the register (right shift).

Referring to the logic diagram in fig (b), when the mode control line is held high, the AND gate on the right input to each NOR gate is enabled while the left AND gate is disabled. The data inputs, A, B, C and D will then be loaded into the register on a negative transition of the clock—this is parallel data input.

When the mode control line is low, the AND gate on the left input to each NOR gate will be enabled and right AND gate will be disabled. The data input to the flip-flop Q_A is now at serial input. On each negative-transition of clock, a data bit is entered serially into the register at the first flip-flop Q_A and each stored data bit is shifted one flip-flop to the right toward the last flip-flop Q_D. This is serial data input and also right shift operation.

In order to effect a shift left operation, the input data must be connected to the D data input as shown in Fig. (c) below.

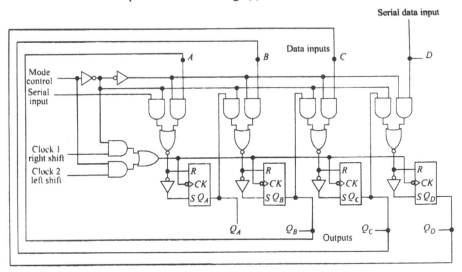

Fig. (c)

It is also necessary to connect Q_D to C, Q_C, to B and Q_B to A. Now when the mode control line is held high, data bit will be entered into flip-flop Q_D and each stored data bit will be shifted one flip-flop to the left on each negative transition of clock. This is also serial input of data (at input (D) and is shift left operation and serial output is at A. For left-shift we find that extra wired connection is needed.

There are two clock inputs - clock 1 and clock 2. This is for the requirement where the clock used to shift data to the right is separate from the clock used to shift data to the left. It is not needed then we can connect clock 1 and clock 2 together. The clock signal will then pass through the AND –OR gate combination non-inverted and the flip-flops will respond to clock negative transitions.

Q.8. Explain Ring Counter.

Ans. Let us consider the serial shift register 74164 which consists of 8 RS flip-flops and that the A and B data inputs are connected together. Now let us connect the output of last flip-flop Q_H back to the data input of the first flip-flop A. Now if we suppose that all the flip-flops are reset and the clock is allowed to run then every time the clock goes high, the zero in each flip-flop will be shifted into the next flip-flop to the right, while 0 in the last flip-flop H will travel along the feedback loop and shift into the first flip-flop A. Since only zeros are moving, one can not appreciate the action much.

In an effort to obtain some action, let us set 1 in the first flip-flop A by using some logic circuit in the feedback path as shown in figure.

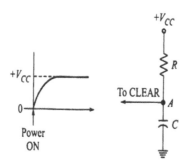

Fig. (a) Power-on-reset circuit

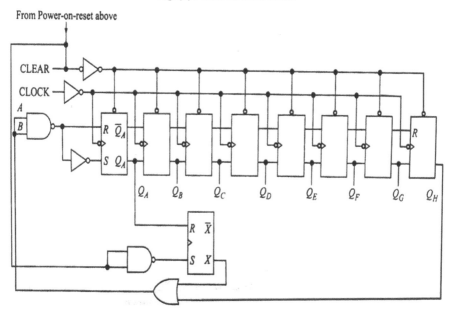

Fig. (b)

The logic added in the feedback path well cause a single 1 to be set into the register in the following manner.

The power-on-reset pulse is inverted and used to initially set flip-flop X. This causes the output of the OR gate to be a 1 and the first clock PT will shift this 1 into Q_A.

When Q_A goes high, this will reset flip-flop X. At this point the register contains an 1 in Q_A and 0's in all other flip-flops. X will remain low as long as power is applied and the data form Q_H will pass through the OR gate directly to the data input AB. The single 1 and the seven 0s will now shift around the register advancing one position with each clock transition. The wave forms are shown below in Fig. (*b*).

Since the 1 moves in a loop form, it appears like a ring and so it is called a Ring counter.

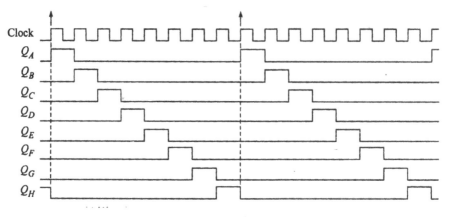

Fig. (*b*) Waveforms

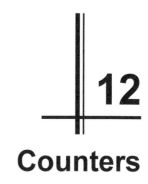

Counters

FACTS THAT MATTER

1. **Counter:** A counter is one of the most useful and versatile subsystems in a digital system. A counter driven by a clock can be used to count the number of clock cycles. Since the clock pulses occur at regular intervals, the counter can be used to measure time and therefore frequency.

2. **Types of counter:** there are basically two types of counter (*i*) asynchronous counter and (*ii*) synchronous counter.

3. **Asynchronous counter:** Is a serial counter in which each flip-flop is triggered by the output of the previous flip-flop and thus the counter has a cumulative settling time. These type of counters are also called ripple counters.

4. **Synchronous counter:** A parallel counter in which all flip-flop are triggered in synchronism with the single clock pulse.

5. **Decoding gate:** A logic gate whose output is high (or low) only during one of the unique states of a counter.

6. **Glitch:** An undesired positive or negative pulse appearing at the output of a logic gate.

7. **Modulus:** The modulus of a counter is the total number of states through which the counter can progress.

8. **Natural count:** A counter has a natural count of 2^n, where n is the number of flip-flops in the counter.

9. **Steering logic:** Counters of any modules can be constructed by incorporating logic which causes certain states to be skipped over or omitted. The technique for skipping counts by steering the clock pulses to certain flip-flops at the proper time is known as steering logic.

10. **Look ahead logic:** In order to omit certain states a technique in which the logic inputs to each flip-flop is preconditioned, is called look-ahead logic.

11. **Lock out of a counter:** counter getting locked into unused states.

12. **Presettable counter:** A counter incorporating logic such that it can be preset to any desired state.

13. **Up Down Counter:** A basic counter, synchronous or asynchronous that is capable of counting in either upward or downward direction.

14. **Sequence Generator:** Generates a binary data sequence.

15. **Decade counter:** A decade counter has 10 states. This is also known as mod-10 counter or divide by 10 counter.

16. **Counters having modified count:** It is often desired to construct counters having a modulus other than natural count like 2, 4, 8 and so on. A counter having modulus 3, 5, 7 etc can always be constructed from a larger modulus counter by skipping states. Such counters are said to have a modified count. The correct number of flip-flops is determined by choosing the lowest natural count that is greater than the desired modified count. For instance, a mod-3 counter requires 2 flip-flops, since 4 is the lowest natural count greater than the modified count of 3.

17. **Higher Modulus counters:** Higher modulus counters can easily be constructed by using combinations of lower-modulus counters. Such configurations are also known as combinational counters. These configurations compromise between speed and hardware count.

18. **Application of Counters:** The most interesting application of counters is in the design of a digital clock.

OBJECTIVE TYPE QUESTIONS

1. How many flip-flops are required to construct a mod-1024 ripple counter.

2. A counter having n flip-flops has modulus of _____.

3. Divide by 8 counter is also called _____.

4. Decade counter is also called _____.

5. What is the largest decimal number representable by a mod-6 ripple counter

6. What is the Boolean expression for an AND gate needed to decode state 6 of a counter.

7. What is the primary cause of glitches that sometimes occur at the output of decoding gate used in ripple counter? How can you eliminate these glitches?

8. How does a parallel counter differ from a serial counter?

9. Why are decoding gate glitches eliminated in a synchronous counter?

10. How many flip-flops are required to construct a mod-12 counter?

Answers

1. Ten.
2. 2^n.
3. Mod-8 counter
4. mod-10 Counter.
5. 63.
6. $C B \overline{A}$ (where A, B,C are inputs and C is MSB)
7. The Primary cause of such glitches is flip-flop propagation time. We can eliminate these by using clock as a "strobe'.
8. In a parallel counter, all the flip-flops change state in synchronism with the single clock pulse whereas in serial counter each flip-flop is triggered by the output of the previous flip-flop.
9. The glitches are eliminated in a synchronous counter or parallel counter because all gate inputs are synchronized that is they are all delayed from the clock by the same amount.
10. Four flip-flops.

SHORT ANSWER TYPE QUESTIONS

Q.1. What is the difference between Asynchronous and synchronous counters?

Ans. The differences between Asynchronous and synchronous counters are as follows:

Asynchronous Counter	Synchronous Counter
(i) Also called serial counter or ripple counter.	(i) Also called parallel counter.
(ii) Each flip-flop is triggered by the output of previous flip-flop, and thus the counter has a cumulative settling time.	(ii) Each flip-flop is triggered by the clock in synchronism and thus the settling time of the counter is equal to the delay time of a single flip-flop.

(Contd...)

Asynchronous Counter	Synchronous Counter
(*iii*) These counters have speed limitation	(*iii*) These counters have increase in speed of operation compared to asynchronous counters.
(*iv*) Its construction requires a minimum of hardware.	(*iv*) Its construction require more hardware compared to asynchronous counter.
(*v*) Glitches may occur at the output of decoding gates used in asynchronous counters.	(*v*) Glitches will not occur at the output of decoding gates used in synchronous counters.

Q.2. What is a presettable counter?

Ans. The presettable counter is the basic building block that can be used to implement a counter that has any modulus.

Nearly all the presettable counters available as TTL MSI are constructed by using four flip-flops and are generally referred to as 4-bit counters. They may be either synchronous or asynchronous. When counted such that the count advances in a natural binary sequence from 0000 to 1111, it is simply referred to as a binary counter. By adding some logic circuit to this binary counter we can preset the counter to certain value. For instance if the preset input is 0110, successive clock pulses produce 0111, 1000 ... reaching a maximum value of output equal to 1111. The next clock pulses resets the count and the data inputs preset the counter to output 0110. Thus here the counter effectively skips states 0 to 5. Thus we get 10 distinct states or counts from 0110 to 1111 which is due to presetting 0110. Thus this counter behaves like mod-10 counter. Again if we change the preset input, we can get different modulus.

LONG ANSWER TYPE QUESTIONS

Q.1. Explain the working of a 3bit binary ripple counter, giving the waveforms.

Ans. A 3 bit binary ripple counter constructed using three negative edge triggered JK flip-flops in cascade is shown in Fig. (*a*) below:

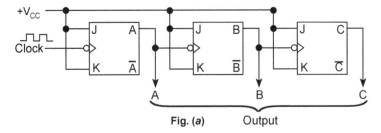

Fig. (*a*)   Output

We see from the Fig. (*a*) that the system clock, a square wave drives flip-flop A. The output of A drives B and output of B drives C. All the J and K inputs are tied to $+V_{CC}$, thus each flip-flop will toggle with a negative transition at its clock input. Since here the output of one flip-flop is acting as the clock input for the next flip-flop, this counter is called asynchronous counter.

The A flip-flop must change state before it can trigger B and B must change state before it can trigger C. The triggers move through the flip-flops like a ripple in water and hence these counters are also known as ripple counters. The overall propagation delay time is the sum of the individual delays. For *e.g.* if each flip-flop has a propagation delay time of 10 *ns*, the overall propagation delay time for the 3 bit ripple counter will be 30 *ns*.

To explain the action or working of the counter let us first draw the waveforms as the clock runs.

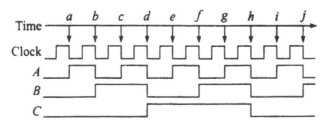

Fig. (*b*) Waveforms

Let us assume that all the flip-flop are initially in reset conditions. If A is the LSB and C is the MSB, the contents of of the counter is CBA = 000.

Now with every clock going negative, A will change state. So from the waveform above at time a, b, c, d and so on, A will change state. If we notice carefully are find that the output of flip-flop A is one half the clock frequency.

Since A acts as the clock for B, with every A going low, B will toggle. Thus at point b, d, f, h and so on, B changes state. Again we find that output of flip-flop B has one-half the frequency of A and one fourth the clock frequency.

Similarly since B acts as clock for C, each time B goes low, C will toggle. Thus C goes high at point d and back to low at point h. The frequency of the waveform at C is one half that of B and one fourth of A and one-eighth of clock frequency.

If we examine carefully, the output condition of the flip-flops is a binary number equivalent to the number of Clock Negative transition that have occurred. Prior to point a on time, output CBA = 000, At point a, CBA changes to 001, at point b, CBA becomes 010 and so on. In fact the counter contents advances one count with each clock NT in a straight binary progression, which is shown in the truth table below:

Negative clock transition	C	B	A	State or count
–	0	0	0	0
a	0	0	1	1
b	0	1	0	2
c	0	1	1	3
d	1	0	0	4
e	1	0	1	5
f	1	1	0	6
g	1	1	1	7
h	0	0	0	0

Truth Table

From the truth table we see that this 3 bit counter can be used to count the number of clock transitions up to a maximum of seven. The counter begins at count 000, advances one count with each NT of clock till it reaches count 111. After this it resets back to 000 and begins the count cycle all over again. So this ripple counter is operating in count up mode.

Moreover here we find that the three flip-flop counter has $2^3 = 8$ output conditions (000 to 111). Thus a three bit binary counter is also referred to as modulus -8 (or mod-8) counter since it has eight states.

Q.2. Draw the wave forms for the 3 bit binary ripple counter shown below and explain its working.

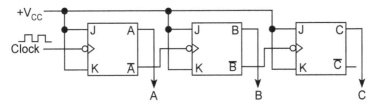

Ans. Figure shows a 3-bit binary ripple counter where flip-flop A is triggered by the system clock input. Complement of output A is used to trigger B and complement of B is used to trigger C. The resulting wave forms are shown below:

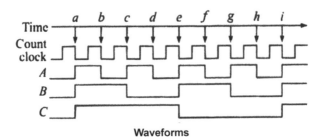

Waveforms

Let us assume that all the flip-flops are initially in reset conditions, Since flip-flop A is triggered by the system clock, A toggles with every clock going negative.

Hence A toggles at point a, b, c, d and so on. Flip-flop B is triggered by the complement of A i.e. $\overline{A}$. So each time $\overline{A}$ goes low or A goes high B will toggle. On the time line, B toggles at points a, c, e, g and i.

Similarly flip-flop C is triggered by $\overline{B}$, so C will toggle each time $\overline{B}$ goes low or B goes high. Thus C toggles at point a, e, and i.

Now if C is MSB and A is LSB, the counter contents become CBA = 111 at point a, CBA = 110 at pt b, CBA = 101 at point C on the time line and so on. Here we find that the counter contents are reduced by one with each clock transition. In other words the counter is operating in a count down mode. The results are summarized in the truth table shown below.

Negative clock transition	C	B	A	State or count
–	0	0	0	0
a	1	1	1	7
b	1	1	0	6
c	1	0	1	5
d	1	0	0	4
e	0	1	1	3
f	0	1	0	2
g	0	0	1	1
h	0	0	0	0
i	1	1	1	7

Truth Table

From the truth table we find that this 3 bit counter can be used to count the number of clock transitions up to a maximum of seven. The counter begins at count 111, reduces one count with each NT of clock till it reaches 000 count. After this it resets back to 111 and begins the count cycle all over again but in count down mode. This is still a mod-8 counter, since it has eight discrete states, but is connected as a down counter.

Q.3. Explain the working of a 3 bit asynchronous up-down counter.

Ans. A 3 bit binary asynchronous up-down counter that counts in a straight binary sequence in up mode and down mode is shown in the figure below:

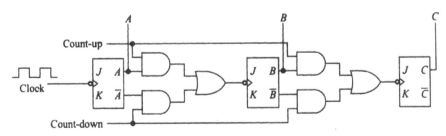

Note: The J and K inputs are all tied to $+V_{CC}$.
The counter outputs are A, B, and C.

The up-down counter is simply the combination of the 3 bit asynchronous up counter and 3 bit asynchronous down counter with some additional gating arrangement.

For this counter to operate in count-up mode, it is necessary that to trigger flip-flop B and C with the true output of A and B respectively, Flip-flop A is triggered by the system clock.

For this counter to operate in count down mode, flip-flop B and C should be triggered by $\overline{A}$ and $\overline{B}$ respectively. Flip-flop A is again triggered by the clock input.

Now from the figure given above we find that if count down control line is low and count up control line is high, the upper AND gates are enabled *i.e.* flip-flop B and C are triggered by A and B respectively. So the counter will act as count-up mode.

(**Note:** The detailed description, count-up wave forms and truth table may be repeated as described in Q. 1 above)

On the other hand if count-down is high and count up control line is low, flip-flop B and C will be triggered by $\overline{A}$ and $\overline{B}$ respectively since the lower AND gate is enabled now. So, the counter will act in count-down mode.

(**Note:** The detailed description, count-down wave-forms and truth table may be repeated as described in Q.2. Above.)

Here one important point is to be noticed that the gates introduce additional delays. So it must be taken into account while determining the maximum rate at which the counter can operate.

Q.4. Explain the working of a 3 bit binary synchronous counter, giving the waveforms.

Ans. A 3 bit binary synchronous or parallel counter constructed using three negative edge-triggered JK flip-flops along with the truth table and the waveforms for the natural count sequence is shown in Fig. (*a*), Fig. (*b*) and Fig. (*c*) below. The J and K inputs of all the flip-flops are tied together to $+V_{CC}$, thus the flip-flop will toggle with a negative transition at its clock input.

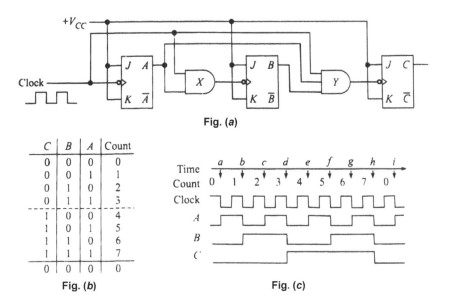

Fig. (a)

C	B	A	Count
0	0	0	0
0	0	1	1
0	1	0	2
0	1	1	3
1	0	0	4
1	0	1	5
1	1	0	6
1	1	1	7
0	0	0	0

Fig. (b)   **Fig. (c)**

From Fig. (*a*) we find that AND gate X is used to gate every second clock to flip-flop B, AND gate Y to gate every fourth clock to flip-flop C. This logic configuration is referred to as "steering logic" since the clock pulses are gated or steered to each individual flip-flop.

The clock is applied directly to flip-flop A. Therefore A will change state with each clock Negative transition, *i.e.* at points *a*, *b*, *c*, *d* and so on, on the time line.

From the Fig. (*a*), we see that whenever A is high, AND gate X is enabled and a clock pulse is passed through the gate to the clock input of flip-flop B. Thus B changes state with every other clock NT or A going negative at points b, d, f and h on time line.

Since AND gate Y is enabled and transmit the clock to flip-flop C only when A and B both are high, flip-flop C changes state with every fourth clock NT at points d and h on time line.

If we examine the waveforms shown in Fig. (*c*) we find CBA = 000 (initially) where C is MSB and A is LSB. At point a, CBA = 001, at b, CBA = 010 and so on, thus advancing one count with each clock NT. This is summarized in truth table in Fig. (*b*). So we find that this counter counts upward in a natural binary sequence from 000 to 111 and again starts the counts from 000. Since there are 8 counts or states, this is a mod-8 parallel or synchronous binary counter operating in count-up mode.

In this parallel counter since all the flip-flops changes states in synchronism, it is not possible to produce a glitch at the output of a decoding gate. Therefore, the decoding gates needed not be strobed as required for asynchronous counters for the possibilities of glitches.

Q.5. Explain the working of a 4 bit synchronous counter, giving the waveforms.

Ans. A 4 bit synchronous or parallel counter constructed using four negative edge-triggered JK flip-flops along with the waveforms and truth table for the natural count sequence is shown below in Fig. (*a*), Fig. (*b*) and Fig. (*c*) The J and K inputs are tied together to $+V_{CC}$ for all the flip-flops, thus the flip-flop will toggle with a negative transition at its clock input.

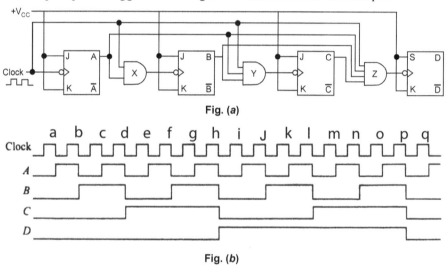

Fig. (*a*)

Fig. (*b*)

From Fig. (*a*) we find that AND gate X used to gate every second clock to flip-flop B, AND gate Y to gate every fourth clock to flip-flop C and AND gate Z to gate every eighth clock to flip-flop D. This configuration is referred to as "steering logic" since the clock pulses are gated or steered to each individual flip-flop.

The clock is applied directly to flip-flop A. Therefore A will change state with each negative transition of the clock i.e. at points a, b, c, d and so on, on the time line.

From Fig. (*a*), we see that whenever A is high, AND gate X is enabled and a clock pulse is passed through the AND gate to the clock input of flip-flop B. Thus B changes state with every other clock NT or A going negative at points b, d, f, h and so on, on time line.

Since AND gate Y is enabled and transmits the clock to flip-flop C only when A and B both are high, flip-flop C changes state with every fourth clock NT at points d, h, I and so on time line or B going negative.

Since AND gate Z is enabled and transmits the clock to flip-flop D only when A, B and C are high, flip-flop D changes state with every eighth clock NT at points h, p on the time line or C going negative.

If we examine the wave forms shown in Fig. (b), we find initially DCBA = 0000 where D is MSB and A is LSB. At point a, DCBA = 0001, at b, DCBA = 0010 and so on, thus advancing one count with each clock NT. This is summarized in the truth table in Fig. (c).

Negative clock transition	D	C	B	A	Count
–	0	0	0	0	0
a	0	0	0	1	1
b	0	0	1	0	2
c	0	0	1	1	3
d	0	1	0	0	4
e	0	1	0	1	5
f	0	1	1	0	6
g	0	1	1	1	7
h	1	0	0	0	8
i	1	0	0	1	9
j	1	0	1	0	10
k	1	0	1	1	11
l	1	1	0	0	12
m	1	1	0	1	13
n	1	1	1	0	14
o	1	1	1	1	15
p	0	0	0	0	0

Fig. (c)

So we see that this 4 bit parallel counter counts upward in a natural binary sequence from 0000 to 1111, and again starts the count from 0000. Since there are 16 counts or states, this is a mod-16 synchronous or parallel counter operating in count up mode.

In this parallel counter since all the flip-flops change states in synchronism, it is not possible for the glitches to occur at the output of a decoding gate. Therefore, the decoding gates need not be strobed as required for asynchronous counter.

Q.6. Design a mod-3 counter and explain its working giving its waveforms.

Ans. Counters having a modulus given by 2^n, where n indicates the number of flip-flops, are said to have a "natural count" of 2^n. Thus mod-2 counter requires a single flip-flop, a mod-4 counter requires two flip-flops, since $2^2 = 4$ and it counts through four discrete states.

Now if we want to construct counter having a modulus other than 2, 4, 8 and so on, it can be constructed from a next larger modulus counter by skipping states. Such counters are said to have a modified count.

It is first necessary to determine the number of flip-flops required. The correct number of flip-flops is determined choosing the lowest natural count that is greater than the desired modified count. So, a mod-3 counter requires two flip-flops, since 4 is the lowest natural count greater than the modified count of 3. One way to design mod-3 counter and its waveforms are shown below:

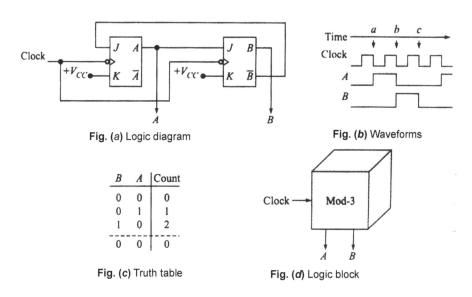

Fig. (a) Logic diagram   Fig. (b) Waveforms

Fig. (c) Truth table   Fig. (d) Logic block

From Fig. (a) we see that K input of both the flip-flops are connected to $+V_{CC}$. $\overline{B}$ of 2nd flip-flop is fed back to J input of 1st flip-flop and A output of 1st flip-flop is connected to J input of 2nd flip-flop. Both the flip-flops are triggered by the NT of clock. Since two flip-flops have a natural count of 4, this counter skips one state. Here's how it works:

(i) Prior to point a on the time line, A = 0, B = 0. A negative clock transition at a will cause flip-flop A to toggle to 1, since J and K inputs are high. B will remain at reset condition i.e. o since J input is low and K is high for B.

(ii) Prior to point b on time line, A = 1, B = 0. Now a negative transition at b will cause A to toggle to 0 since its J and K inputs are high B will now also toggle to 1 since its J and K inputs are high.

(*iii*) Prior to point C on time line, A = 0, B = 1 A NT of clock at C will cause A to remain at 0 since its J input is low K input is high. B to reset to 0 since its J input is low and K is high.

(*iv*) The counter thus progressed through three states advancing one count with negative clock transition. This is shown in the truth table in Fig. (*c*).

This two flip-flop mod-3 counter can be considered as a logic building block as shown in Fig. (*d*). It has a clock input and outputs at A and B. It can be considered as a divide-by-3 block, since the output waveform at B or at A has a period equal to three times that of the clock. In other words this counter divides the clock frequency by 3.

We also find that this is a synchronous counter since both flip-flop change state in synchronism with the clock.

Q.7. Design a mod-6 counter and explain its working giving its waveforms.

Ans. If we consider basic flip-flop to be a mod-2 counter, we see mod-4 counter is 2 × 2 connection *i.e.* two mod-2 counters in series. Similarly mod-8 counter is simply 2 × 2 × 2 connection.

Thus if we connect a flip-flop at the output B of the mod-3 counter, the result is (3 × 2 = 6) mod-6 counter as shown below in Fig. (*a*) and waveforms in Fig. (*b*).

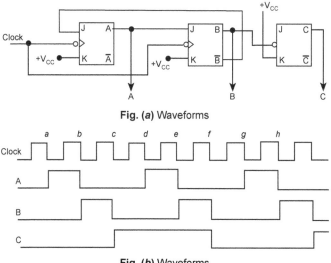

Fig. (a) Waveforms

Fig. (b) Waveforms

From Fig. (*a*) are see that K input of all the three flip-flops are connected to +V$_{CC}$. $\overline{B}$ of 2nd flip-flop is fed back to J input of 1st flip-flop and A output of 1st flip-flop is connected to J input of B flip-flop. Both the flip-flops are triggered by the NT of clock J input of flip-flop C is also connected to $+V_{CC}$ and this flip-flop is triggered by B. Now if we examine the waveforms, we find the counter works as follows.

(*i*) Prior to point a on time line, A = 0, B = 0, C = 0. A negative clock transition at a will cause flip-flop A to toggle to1, since J and K inputs are high. B and C will remain at reset condition *i.e.* 0 since J input is low and K is high for B and B (which is acting as clock for C) is low so C will also remain 0.

(*ii*) Prior to point b on time line, A = 1, B = 0, C = 0. Now a negative transition at pt b will cause A to toggle to 0 since its J and K inputs are high. B will now also toggle to 1 since its J and K inputs are high C will remain 0, since B (clock for C) was low.

(*iii*) Prior to pt-C on time, A = 0, B = 1, C = 0. A NT of clock at C will cause A to remain at 0 since its J input is low and K input is high. B will reset to 0 since J input for B is low and K is high. C will now toggle to 1 since B (clock) for C was high.

Thus if we go on examining in the same manner, we will find that further A will change state at d, e, g and so on. B will further change state at e, f, h and so on and C will change with every B going negative. Thus the truth table can be found out as given below:

Negative transition	C	B	A	Count
–	0	0	0	0
a	0	0	1	1
b	0	1	0	2
c	1	0	0	4
d	1	0	1	5
e	1	1	0	6
f	0	0	0	0
g	0	0	1	1

Here C is MSB and A is LSB. Thus we find that this counter has 6 counts or states (0, 1, 2, 4, 5, 6). Again it repeats the same counts. Count 3 and count 7 are missing in this kind of arrangement, but since we are getting 6 states this counter is known as mod-6 counter.

Also, this counter is not a synchronous counter since flip-flop C is triggered by flip-flop B, that is the flip-flops do not all change in synchronism with the clock. This counter is not even asynchronous counter since flip-flop A and B are both triggered by the clock. This kind of counter is known as combination counter.

Q.8. Draw the waveforms and explain the working of the mod-6 counter by connecting a single flip-flop in front of mod-3 counter.

Ans.

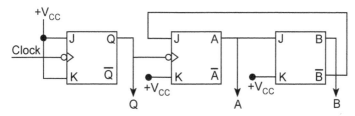

Fig. (*a*) shows that a single flip-flop Q is added in front of mod-3 counter, the result is (2 × 3 = 6) mod-6 counter. Let us now draw the Waveforms and explain the working.

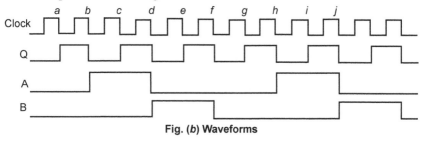

Fig. (*b*) Waveforms

(*i*) Prior to point a on time line, Q = 0, A = 0, B = 0. A negative clock transition at a will cause flip-flop Q to toggle to1, since J and K inputs are high. A and B will remain at reset condition *i.e.* 0 since J input for A is high and for B is low and K input is high for both but clock Q for A and B is low.

(*ii*) Prior to point *b* on time line, Q = 1, A = 0, B = 0. A –ve clock transition at *b* will cause Q to toggle to 0. A will toggle to 1 since Q (clock for A is going –ve) and J and K for A both are high, B will remain 0 since J is low and K is high.

(*iii*) Prior to point *c* on time line, Q = 0, A = 1, B = 0. A negative clock transition at *a* will cause clip-flop Q to toggle to 1, since J and K inputs are high. A will remain as 1 since Q (clock) was 0 and B will also remain at 0 since clock (Q) for B is 0.

Thus if we examine the waveforms we get the following truth table:

Negative transition	B	A	Q	Count
–	0	0	0	0
a	0	0	1	1
b	0	1	0	2
c	0	1	1	3
d	1	0	0	4
e	1	0	1	5
f	0	0	0	0

Here B is MSB and Q is LSB. Thus we find that this counter has 6 counts or states (0, 1, 2, 3, 4, 5). Again it repeats the some counts. So if we connect (2 × 3 = 6), we get count from 0 to 5 and the counter skips count 6 and count 7. Thus in this arrangement the counter advances one count at a time starting from 0 to 5 and thus it is a mod-6 counter.

Also this counter is neither an asynchronous counter nor a symmetrical counter because here Q is triggered by the clock but A and B are triggered by Q acting as clock. So this is a combination counter.

Q.9. Design a mod-5 counter and explain its working giving its waveforms.

Ans. For a mod-5 counter we need three flip-flops. The required design for mod-5 counter which will skip three counts and advance one count at a time, through strict binary sequence beginning from 000 to 100 is shown below with waveforms and truth table.

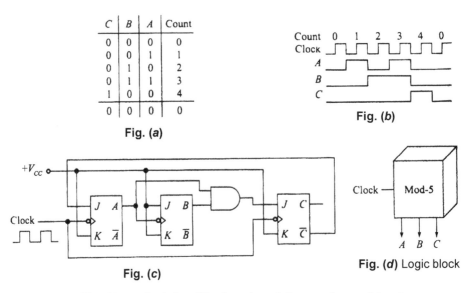

C	B	A	Count
0	0	0	0
0	0	1	1
0	1	0	2
0	1	1	3
1	0	0	4
0	0	0	0

Fig. (a)

Fig. (b)

Fig. (c)

Fig. (d) Logic block

From Fig. (c) we find that flip-flop A and C are triggered by the system clock and flip-flop B is triggered by A. Inputs J and K of flip-flops B are tied to $+V_{CC}$. Input K of A and C is also connected to $+V_{CC}$. The output of the AND gate whose inputs are A and B is the input J for flip-flop C. $\overline{C}$ is the input J for flip-flop A.

So A will change state whenever clock goes negative and $\overline{C}$ is high. Flip-flop B will change state whenever A goes negative or with every negative transition of A.

Flip-flop C will change state when A and B both are high and clock goes negative.

If we notice the waveforms in Fig. (b) we see that $\overline{C}$ is high during all counts except count 4. Now if $\overline{C}$ is connected to J input of flip-flop A we will have the desired inhibit signal during count 4 such that we have count 0 after count 4. This is true since the J and K inputs to flip-flop A are both high for all counts except count 4. So flip-flop A toggles each time the clock goes negative, but during count 4, the J side is low and with next NT of clock, flip flop A remains at reset condition or 0.

Now for flip-flop C, J input is high only during count 3, C will be high during count 4 and low during all other counts. At all other times the J input to flip-flop C is low and is hold in the reset state.

Thus we get the truth table as shown in Fig. (*a*) which has five counts from 0 to 4 and skips count 5, 6 and 7. Thus it is a mod-5 counter. This is a combination counter.

Mod-5 can be used as logic block, as shown in Fig. (*d*) and can be used in cascade to construct higher modulus counter. For *e.g.* a 2 × 5 or 5 × 2 will form a mod-10 counter or decade counter.

Q.10. Show that the mod-5 counter which skips count 5,6 and 7 does not malfunction.

Ans. The logic circuit using three flip-flops which skips count 5, 6 and 7 and acts as mod-5 counter having counts from 0 to 4 is shown below:

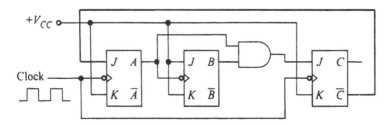

Logic Circuit

Let us examine the omitted states to make sure that the counter will not malfunction. This counter omits states 5, 6 and 7 during its normal operating sequence. There is however a possibility that the counter may set up in one of the omitted or illegal states when power is first applied to the system. Now if with the next clock going negative, the counter progresses into the normal count sequence, we can say that the counter does not malfunction. Let us assume that as soon as the power is on, the counter is in state 5 *i.e.*, CBA = 101. When the next clock pulse goes low

(*i*) Since $\overline{C}$ is low, flip-flop A resets. Thus A changes from 1 to 0.

(*ii*) Since A changes from 1 to 0, B triggers and B changes from 0 to 1.

(*iii*) Since J input of flip-flop C is low, C changes from 1 to 0.

Thus the counter progresses from illegal state 5 to the legal state 2 (CBA = 010).

Next let us assume that the counter starts in the illegal state 6 (CBA = 110). When the next clock pulse goes low:

(*i*) Since $\overline{C}$ is low, flip-flop A resets. A is already 0, it just remains 0

(*ii*) Since A does not change, B does not change and is remains at 1

(*iii*) Since J input of flip-flop C is low, flip-flop C resets and C changes from 1 to 0.

Thus the counter progresses from the illegal state 6 (CBA = 110) to legal state 2 (CBA = 010).

Finally let us assume that the counter is in illegal state 7 (CBA = 111). When the next clock pulse goes low:

(*i*) Since $\overline{C}$ is low, flip-flop A resets and A changes from 1 to 0

(*ii*) Since A changes from 1 to0, B changes from 1 to 0

(*iii*) The J input to flip-flop C is high, therefore flip-flop C toggles from 1 to 0.

Thus the counter progresses from the illegal state 7 to legal state 0 (CBA = 000).

Thus we find that none of the three illegal states cause the counter to malfunction.

Q.11. Design a decade counter (mod-10) and explain its working giving its waveforms.

Ans. If we consider basic flip-flop to be a mod-2 counter and connect it at the output of C of the mod-5 counter, the result is (5 × 2 = 10) mod-10 counter as shown below in Fig. (*a*) waveforms in Fig. (*c*) and truth table in Fig. (*b*).

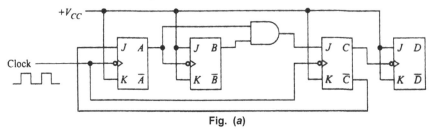

Fig. (a)

D	C	B	A	State
0	0	0	0	0
0	0	0	1	1
0	0	1	0	2
0	0	1	1	3
0	1	0	0	4
1	0	0	0	5
1	0	0	1	6
1	0	1	0	7
1	0	1	1	8
1	1	0	0	9
0	0	0	0	0

Fig. (b)

Fig. (c)

From Fig. (a) we find that flip-flop A and C are triggered by the system clock and flip-flop B is triggered by A and flip-flop D is triggered by C. Input J and K of flip-flop B and D are tied to $+V_{CC}$. Input K of flip-flop A and C are also connected to $+V_{CC}$. $\overline{C}$ is connected to J input of flip-flop A and the output of AND gate whose inputs are A and B is the input J for flip-flop C. So A will change state whenever $\overline{C}$ is high and clock goes negative.

Flip-flop B will toggle with every A going negative.

Flip-flop C will change state when A and B both are high and clock goes negative.

Flip-flop D will change state with every C going negative.

If we notice the waveforms in Fig. (c), we see that $\overline{C}$ is high during all counts except count 4 and 9. Since $\overline{C}$ is connected to J input of flip-flop A, during count 4 and 9, the J side is low and with next NT of clock, flip-flop A remains at reset condition or 0.

Now for flip-flop C, J input is high only during count 3 and 8, so C will be high during count 4 and 9 and low during all other counts. At all other time, the J input to flip-flop C is low and is held in the reset state.

Thus we get the truth table as shown in Fig. (b). If we notice properly, we find the counter progresses through a biquinary count sequence and does not count in a straight binary sequence. But this counter has ten discrete states or counts, so it is a mod-10 counter or decade counter.

Q.12. Design a decade counter which counts in a straight binary sequence and explain its working giving its waveforms.

Ans. If we connect a single flip-flop in front of mod-5 counter, the result is $(2 \times 5 = 10)$ 10 counter which counts in a straight binary sequence from 0000 to 1001 and back to 0000. The truth table, waveforms and the decade counter connected in 2×5 configuration are shown (a), (b) and (c) below:

D	C	B	A	Count
0	0	0	0	0
0	0	0	1	1
0	0	1	0	2
0	0	1	1	3
0	1	0	0	4
0	1	0	1	5
0	1	1	0	6
0	1	1	1	7
1	0	0	0	8
1	0	0	1	9.
0	0	0	0	0

Fig. (a)

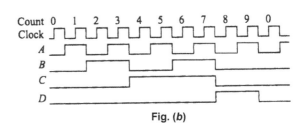

Fig. (b)

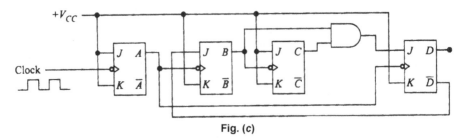

Fig. (c)

From Fig. (c) we find that flip-flop A is triggered by the system clock. Flip-flop B and D are triggered by A and flip-flop C is triggered by B. J and K inputs of flip-flop A and C are connected to $+V_{CC}$. So flip-flop A will toggle with every clock going negative flip-flop C will toggle with every B going negative.

K inputs of flip-flop B and D is connected to $+V_{CC}$. $\overline{D}$ is connected to J input of flip-flop B and the output of the AND gate whose inputs are B and C, is connected to J input of flip-flop D. So flip-flop D will change state on the next transition of flip-flop A when B and C was high i.e. D will change from 0 to 1 during count 8 as we see from the wave form and remain as 1 during count 9 also as A is going from 0 to 1 during count 9. With the next NT of A, D becomes 0.

Flip-flop B will change state whenever A goes negative and $\overline{D}$ is high. $\overline{D}$ is high during all counts except count 8 and 9.

Thus we get the resulting waveforms as shown in Fig. (b) above and the corresponding truth table as shown in Fig. (a). If we examine the truth table we find that the counter progresses through straight binary sequence from 0000 to 1001 i.e. count 0 to 9 and back to 0. Since it has ten discrete state or counts, it is a mod-10 counter or decade counter.

Q.13. Design a binary counter using T flip-flops, the counting states of which are shown in state transition diagram.

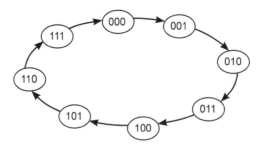

Ans. To design the binary counter the first step is to get the state table from the given sequence as shown in figure below:

Present state			Next state			Flip-flop inputs		
A	B	C	A^+	B^+	C^+	T_A	T_B	T_C
0	0	0	0	0	1	0	0	1
0	0	1	0	1	0	0	1	1
0	1	0	0	1	1	0	0	1
0	1	1	1	0	0	1	1	1
1	0	0	1	0	1	0	0	1
1	0	1	1	1	0	0	1	1
1	1	0	1	1	1	0	0	1
1	1	1	0	0	0	1	1	1

State Table for binary counter

The state table shows the present state of flip-flops A, B and C (before a clock pulse is received). Flip-flops change state after a pulse is received which is shown in the table as A^+, B^+ and C^+. The third column in the table is used to derive the inputs for T_A, T_B and T_C.

Whenever the entries in the C and C^+ columns differ, flip-flop C must change state and T_C must be 1. Similarly if B and B^+ differ, B must change state so T_B must be 1, whenever A and A^+ differ, T_A must be 1. For example if ABC = 011, $A^+B^+C^+$ = 100, then all the flip-flops have changed states so, $T_AT_BT_C$ will be equal to 111.

Next step is to find T_A, T_B and T_C as functions of A, B and C from the state table. By observing we find $T_C = 1$.

To find the values of T_A and T_B Karnaugh maps are to be plotted for T_A and T_B separately as follows:

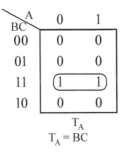

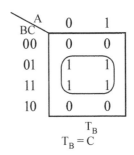

$T_A = BC$ $T_B = C$

Since clock pulse P is required to initiate each change of state, each input has to be multiplied by P.

Hence $T_A = BCP$, $T_B = CP$, $T_C = P$

From these equation, the required binary counter using T flip-flops is designed as follows:

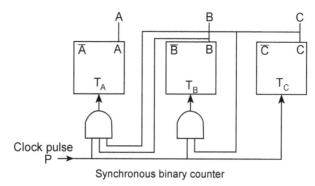

Synchronous binary counter

Q.14. Design a binary counter using JK flip-flops, the counting sequence of which are shown in state transition diagram.

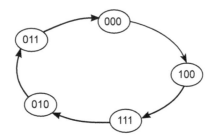

Ans. The given state graph shows that the sequence of states of the counter is not in straight binary order. The arrows indicate the state sequence starting from state 000. It is implicitly understood that the change of state will occur with the clock pulse.

To design the binary counter input equation for J and K of each flip-flop must be derived. Table (*a*) below gives the next state Q^+ as a function of J, K and Q.

J	K	Q	Q⁺
0	0	0	0
0	0	1	1
0	1	0	0
0	1	1	0
1	0	0	1
1	0	1	1
1	1	0	1
1	1	1	0

Table (*a*)

Q	Q⁺	J	K
0	0	0	0
		0	1
0	1	1	0
		1	1
1	0	0	1
		1	1
1	1	0	0
		1	0

Table (*b*)

Q	Q⁺	J	K
0	0	0	X
0	1	1	X
1	0	X	1
1	1	X	0

Table (*c*)

Using Table (*a*) we can derive the input conditions required for J and K when Q and Q⁺ are given. Thus if a change from Q = 0 to Q⁺ = 1 is required, either the flip-flop can be set to 1 by using J = 1 and K = 0 or the state can be changed by using $J = K = 1$ or J must be 1 but K is don't care. Similarly if a change from $Q = 1$ to $Q^+ = 0$ is required, J must be don't care and K must be1. When no state change is required, the inputs J and K are either 0 and don't care or don't care and 0. The J-K input requirements are summarized in Table (b) and Table (c) above.

Now we will write the state transition Table for the given state graph with columns added for the J and K flip-flop inputs. The J and K input columns will be filled using Table (c) above.

A	B	C	A^+	B^+	C^+	J_A	K_A	J_B	K_B	J_C	K_C
0	0	0	1	0	0	1	X	0	X	0	X
0	0	1	–	–	–	X	X	X	X	X	X
0	1	0	0	1	1	0	X	X	0	1	X
0	1	1	0	0	0	0	X	X	1	X	1
1	0	0	1	1	1	X	0	1	X	1	X
1	0	1	–	–	–	X	X	X	X	X	X
1	1	0	–	–	–	X	X	X	X	X	X
1	1	1	0	1	0	X	1	0	X	X	1

For ABC = 000, A = 0 and A^+ = 1 so J_A = 1 and K_A = X as shown in table (c) above. For ABC = 010, A = 0 and A^+ = 0, so J_A = 0 and K_A = X. Similarly for ABC=000, B = 0 and B^+ = 0, so J_B = 0 and K_B = X. For ABC = 011, C = 1 and C^+ = 0, so J_C = X and K_C = 1. The remaining values of J and K for each flip-flop are filled in the same manner. The resulting J-K input function are now plotted as K-maps.

$$J_A = B'$$ $$K_A = C$$ $$J_B = A$$ $$K_B = A'C$$ $$J_C = A+B$$ $$K_C = 1$$

After deriving the flip-flop input equations from the Karnaugh maps, we can design the binary counter as follows.

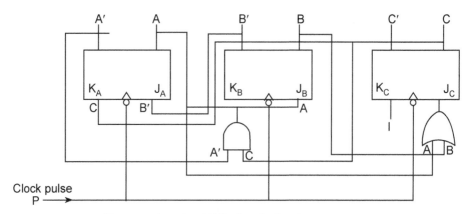

Binary counter using J-K flip-flops for the given sequence.

It is to be mentioned here that since clocked J-K flip-flops have been used, the input pulses go directly to the clock inputs on the flip-flops.

13

Semiconductor Memories

1. **Semiconductor memory:** It consists of a rectangular array of memory cells, fabricated on a silicon wafer and housed in a convenient package, such as DIP.

2. **Memory Cell:** The memory cell is typically a transistor flip-flop or a circuit used to store a single bit of information in a semiconductor memory chip.

3. **Chip:** Chip is a term commonly used to refer to a semiconductor circuit on a single silicon die.

4. **Capacity of memory:** The total number of cells in a memory determine its capacity e.g. a 1024 bipolar memory chip is a semiconductor memory that has 1024 memory cells, each cell consisting of a flip-flop constructed with the use of bipolar transistors.

5. **Address:** Selection of a cell in a memory array for a read or a write operation.

6. **Matrix addressing:** Selection of a single cell in a rectangular array of cells by choosing the intersection of a single row and a single column.

7. **Memory Bank:** Connects more than one memory block to increase the capacity of the memory.

8. **Nonvolatile storage:** A method whereby a loss of power will not result in a loss of stored data.

9. **Flash Memory:** A kind of nonvolatile memory which can be written and erased electrically.

10. **Volatile storage:** A method of storing information whereby a loss of power will result in a loss of the data stored.

11. **Static Memory:** A memory capable of storing data indefinitely, provided there is no loss of power.

12. **Dynamic memory:** A memory whose contents must be restored periodically.

13. **Pass transistor:** A MOS transistor that passes information in either direction when it is turned on.

14. **Floppy disk:** A movable low capacity magnetic storage media.

15. **Hard disk:** A high capacity magnetic storage media, integral part of modern computer.

16. **Read operation:** The act of detecting the contents of a memory.

17. **Write operation:** The act of storing information in a memory.

18. **ROM:** Read only memory

19. **RAM:** Random Access memory

20. **SRAM:** Static RAM

21. **DRAM:** Dynamic RAM

22. **PROM:** Programmable Read only memory

23. **EPROM:** Erasable programmable read only memory

24. **EEPROM:** Electrically Erasable Programmable Read Only Memory

25. **CD-ROM:** Compact Disk Read Only memory, a kind of movable optical storage media that has higher capacity compared to magnetic counter part.

26. **CD-R:** Compact Disk Recordable, a kind of optical memory on which data can be written but once.

27. **CD-RW:** Compact Disk Rewritable, a kind of optical memory on which data can be written and erased many times.

28. **Cache:** It is a small, fast SRAM used to momentarily store selected data in order to improve computer speed of operation.

29. **Classification of semiconductor memories:** Memories are usually classified as i) bipolar, ii) metal oxide semiconductor (MOS) ii) Complementary metal oxide semiconductor (CMOS) according to the type of transistor used to construct the individual memory cell. SRAMs can be either bipolar or MOS, but all DRAMs are MOS

30. **Mask Programmable:** It refers to PROM that can be programmed only by the manufacturer.

31. **Packing Density:** Number of memory bits packed in per unit space.

32. **CAM:** Content Addressable Memory.

33. The two common categories of memory RAM and ROM can be further divided as shown below:

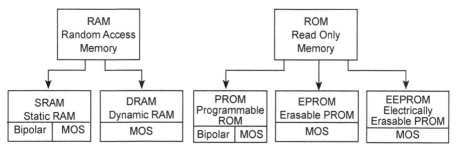

34. **RAM:** Block diagram of a RAM chip is shown in the fig below. An application in which data changes frequently, we use RAM. The logic circuit of RAM allows a single bit of information to be stored in any of the memory cells—this is write operation. There is also logic circuitry that will detect whether a 0 or a 1 is stored in any particular cell—this is read operation. Since a bit can be written (stored) in any cell or read (detected) from any cell, this is known as Random Access memory. A control signal called chip-select or chip-enable, is used to enable or disable the chip. The address lines determine the cells written into or read from. Since each cell is a transistor circuit, a loss of dc power means a loss of data. A RAM that has this type of memory cell is said to provide volatile storage.

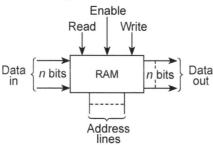

A dc power supply is required to energize any semiconductor memory chip. Once dc power is applied to a static RAM, the SRAM retains stored information indefinitely, without any further action. A dynamic RAM on the other hand does not retain stored data indefinitely. Any stored data must be stored again or refreshed periodically. Both SRAMs and DRAMs are used to construct the memory inside a microcomputer or minicomputer. DRAMs are used as the bulk of memory and high speed SRAMs are used for a smaller, rapid access type of memory known as cache memory. The cache is used to momentarily store selected data in order to improve computer speed of operation. SRAMs can be either bipolar or MOS, but all DRAMs are MOS.

35. **ROM:** Block diagram for ROM chip is shown below:

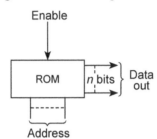

The information or data stored in a ROM is fixed and will be retained permanently even if dc power is removed. Therefore a ROM is ideal for storing permanent instructions necessary for the startup and operation of a computer. Hence instructions like the values of mathematical constants such as trigonometric functions or a fixed program such as that used to find the square root of a number etc could be stored in a ROM. These instructions are retained even when the computer is off and become immediately available each time the computer is turned on. The content of ROM is fixed during manufacturing. A ROM is still random access since there are logic circuitry and address lines to select any desired cell in the memory. When enabled, data from the selected cells is made available at the output. There is of course, no write mode since data is permanently stored in each cell, a loss of power does not cause a loss of data and thus a ROM provides nonvolatile data storage.

Data stored in a programmable ROM (PROM) is permanent—a PROM can be programmed only once.

Data stored in an erasable PROM (EPROM) can be erased. The EPROM can then be used to store new Data. PROMs can be either bipolar or MOS, but all EPROMs are MOS.

The process of entering information into a ROM is referred to as programming the ROM.

Depending on the programming process employed, the ROM's are classified as

    (*i*) **Mask programmable ROM (MPROM):** In these memories the data pattern must be programmed as part of the fabrication process. Once programmed the data pattern can never be changed. They are highly suited for very high volume usage due to their low cost.

    (*ii*) **Programmable ROM (PROM):** A PROM is electrically programmable i.e. the data pattern is defined after final packaging rather than when the device is fabricated. The programming is done with equipment referred to as PROM programmer.

(*iii*) **Erasable PROM (EPROM):** In these memories data can be written any number of times *i.e.* they are re-programmable. For erasing the contents of the memory one of the following methods are employed:

(*a*) Exposing the chip to UV radiation for about 30 minutes.

(*b*) Erasing electrically by applying voltage of proper polarity and amplitude. Electrically erasable PROM is also referred to as E^2 PROM or EEPROM. Re-programmable ROMs are possible only in MOS technology.

OBJECTIVE TYPE QUESTIONS

1. What is memory cell?
2. SRAM stands for _____.
3. DRAM stands for _____.
4. What is the operational difference between an SRAM and a DRAM?
5. What is EPROM?
6. What is a cache memory?
7. The process of entering information into a ROM is known as _____.
8. SRAMs can be either bipolar or MOS, but all DRAMs are _____.
9. EPROMs are possible only in _____ technology.
10. What is the important similarity between BJT based memory cell and CMOS based memory cell?
11. What is access time in memory read operation?
12. What does it mean to say that a chip is mask-programmable?
13. What is DVD?
14. What is the capacity of a single sided double layer DVD ROM?
15. CAM stands for _____.

Answers

1. A transistor flip-flop used to store a single bit of information in a semiconductor memory chip.
2. Static Random Access Memory.
3. Dynamic Random Access Memory.
4. A DRAM must be refreshed periodically which is not required for SRAM.
5. EPROM stands for erasable–programmable read only memory.
6. Cache memory is a small high speed SRAM used to momentarily store selected data in order to improve computer speed of operation.
7. Programming the ROM.
8. MOS.

9. MOS.

10. Both use cross-coupled transistor to store binary value.

11. Time required to get proper data, after the address is stabilized.

12. It refers to a ROM whose contents are established during the manufacturing process.

13. DVD stands for Digital Versatile Disk or Digital Video Disk.

14. 8.5 GB.

15. CAM stands for Content Addressable Memory.

LONG ANSWER TYPE QUESTIONS

Q.1. Explain the semiconductor memory organization. Give example.

Ans. A digital processing system requires a facility for storing digital information. The information usually consists of instructions (processing steps) coded in binary form, data to be processed, intermediate and final results etc. The subsystem of this digital processing system which provides the storage facility is known as memory. Previously the memories used were of magnetic type. With the development of semiconductor technology it has now become possible to make semiconductor memories of various types and sizes. These memories are of small size, low cost, high speed, high reliability and ease of expansion of the memory size. The basic element of a semiconductor memory is a flip-flop. The information is stored in binary form. There are a number of locations in a memory chip. Each location is meant for one word of digital information. The number of locations and number of bits comprising the words vary from memory to memory. The size of a memory chip is specified by MXN bits where M is number of locations and N is the number of bits at each location. Hence M words of N bits each can be stored in the memory. The block diagram of memory is shown below:

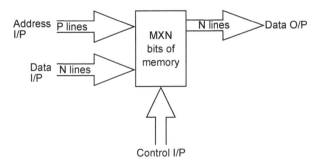

A word is always treated as an entity and moves in and out of memory as one unit. For controlling the movement of these words in and out of memory two signals write and read are used respectively. The words to be written in and also words to be taken out of memory are first entered in a register called memory buffer register (MBR). The location in the memory unit where a word is stored is called address of the word. All the addresses of various words are written in a specific register called memory address register (MAR).

Each M locations in the memory is defined by a unique address and therefore for accessing any one of the M locations or addresses, p inputs are required where $2^p = M$. This set of lines is referred to as address I/Ps or Address Bus. The address is specified in binary form. Generally octal and Hexadecimal representations are commonly used. The address I/Ps applied to a p- to M-decoder circuit which activates one of its M outputs depending on the address and thus the desired memory location is selected. For instance if $M = 16$, $2^p = M$ ∴ $p = 4$. The number of inputs required to store the data into or read the data from each memory location is say N. One set of N lines is required for storing the data into the memory referred to as data I/P and another set of N lines is required for reading the data already stored in the memory referred to as Data O/P. But in some memory chips the same set of N lines is used for Data I/P and Data O/P and this is referred to as Data Bus.

Control I/Ps are required to give command to the device to perform the desired operation either read or write Example of 16×4 memory chip.

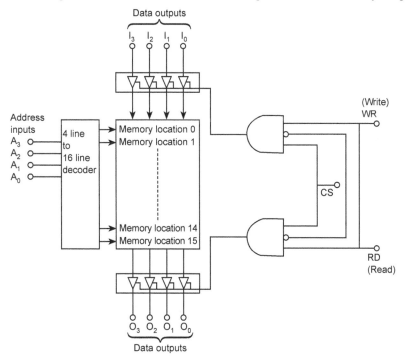

Internal structure of 16 × 4 memory chip

Q.2. Explain the Read and Write operation in a memory.

Ans. Fig below shows the functional block diagram for Read and Write operation in a memory.

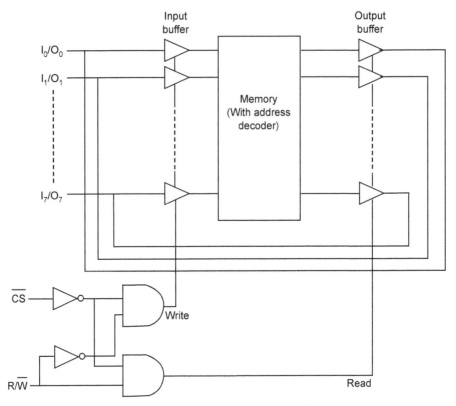

Fig. Memory with common I/P/o/p ports and control logic for read-write

In the above diagram we find there are two control inputs Read-Write, $R/\overline{W}$ and chip-select $\overline{CS}$. The control logic is such that if $\overline{CS} = 1$ then both Read and Write control outputs are 0 which takes both input and output buffer to high impedance state. The memory block/chip is effectively not selected. If $\overline{CS} = 0$ and $R/\overline{W} = 1$, then Read = 1 and WRITE = 0. This time Data pins acts as output (O_7 .... O_0) and memory is read as in this case the output buffer is enabled.

Therefore to read or retrieve a data word stored at a particular address, the following sequence of operations is required to be performed :-

(*i*) The chip-select signal $\overline{CS}$ should be made 0

(*ii*) The address of the desired memory location is applied to the address input lines.

(*iii*) A READ command signal is applied to the READ control I/P

In response to the above operations the data word stored at the addressed location appears in the data output lines.

If $\overline{CS}$ = 0 and R/$\overline{W}$ = 0, Then Read=0 and WRITE=1. This time Data pins acts as input (I_7 .... I_1) as this combination enables the input buffer so that the word gets written to the memory.

Therefore for writing a word into a particular memory location, the following sequence of operation is to be performed:

(*i*) The chip select signal $\overline{CS}$ should be made 0

(*ii*) The word to be stored is applied to the data input lines

(*iii*) The address of the desired memory location is applied to the address input lines

(*iv*) A write command signal is applied to the WR control input.

In response to the above operation the addressed memory location is cleared of any word that might have been stored in it and the information presented at the data I/P terminal replaces it.

Q.3. How can you classify various memory devices?

Ans. Various memory devices can be classified on (a) the basis of principle of operation, (b) physical property, (c) mode of access and (d) fabrication technology.

(*a*) According to the principle of operation memories are classified as

(*i*) Sequentially accessed memory

(*ii*) Read and write memory (RAM)

(*iii*) Read only memory (ROM)

(*iv*) Content addressable memory (CAM)

There are two types of sequentially accessed memories:

(*i*) charged coupled devices (CCD)

(*ii*) shift Register

In a sequentially accessed memory the memory locations are accessed for writing into or reading from in a sequential manner. Therefore the access time for writing into or reading from is different for different location.

(*iii*) **RAM:** Read and write memory is a Random access memory. Here the access time is same for every location. The RAMs can be static or dynamic.

(*iv*) **ROM:** ROM is meant only for reading the information from it. This does not mean that the information is not written into it, because unless any information is stored in it there cannot be anything to read from. The process of entering information into this type of memory is much more complicated than RAM and it is done outside the system where it is used and therefore it is called ROM.

(*v*) **CAM:** CAM is a special purpose RANDOM Access memory which performs association operation in addition to read or write operation.

According to physical property or characteristic memory can be classified as

(*i*) Erasable an erasable memory

(*ii*) Volatile and non-volatile memory.

(*i*) **Erasable memory:** A memory in which the information stored can be erased and new information stored is called erasable memory. Erasable memory can be sub-classified as:

(*a*) Location by location erasable memory in which the desired memory location can be erased one by one and new information can be entered, e.g. Electrically alterable ROM (EAROM).

(*b*) All memory locations are simultaneously erasable in which the contents of all the locations of the memory chip get erased simultaneously by exposing the memory chip to UV radiation. Such a memory is known as Erasable Programmable ROM (EPROM)

(*ii*) **Volatile memory:** If the information stored in a memory is lost when the electrical power is switched off, the memory is referred to as a volatile memory e.g. RAM.

(*iii*) **Non-Volatile memory:** In this type of memory the information once stored remains intact until changed deliberately. All types of ROMs are non-volatile memories.

(*c*) On the basis of fabrication technology the memories can be classified as Bipolar and unipolar.

Static RAM, ROM and PROM can be fabricated either using bipolar technology or MOS technology whereas dynamic RAM, EPROM and EAROM can be fabricated only using unipolar devices *i.e.* MOS technology.

Q.4. How can you expand word length and word capacity of a memory?

Ans. Expanding word length:

Let us assume that the desired word size = n

Word size of available memory chip = N where $n > N$

Number of I.C chips required is an integer next higher to the value $\frac{n}{N}$. These chips are connected in the following manner:

(*i*) Connect the corresponding address lines of each chip individually *i.e.* A_0 of each chip is connected together and it becomes A_0 of the overall memory. Similarly connect other address lines.

(*ii*) Connect the RD input of each IC together and it becomes the RD input for the overall memory. Similarly connect WR and CS inputs.

Now the number or data I/P/ O/P lines will be equal to the product of the number of chips used and the word size of each chip. Thus the word length of the memory is expanded.

Expanding Word Capacity

For expanding word capacity the chips should be connected in the following manner:

(*i*) Connect the corresponding address lines of each chip individually.

(*ii*) Connect the RD I/P of each chip together similarly connect the WR inputs.

(*iii*) Use a decoder of proper size and connect each of its output to one of the CS terminals of memory chip. For *e.g.* if 8 chips are to be connected, a 3 line to 8 line decoder is required to select one out of 8 chips at any time.

Thus the word capacity of the memory chip is expanded.

Q.5. What are the differences between static RAM and dynamic RAM?

Ans. MOS devices use two different techniques to store information depending on the type of basic memory cell. It can be categorized as either static RAM or Dynamic RAM.

Differences between Static RAM and Dynamic RAM are as follows:

Static RAM	Dynamic RAM
1. Static MOS memories are slower	1. Dynamic memories are faster.
2. Easier to drive than dynamic memories .	2. Difficult to drive than static memories since they require clock signal in addition to power supply.
3. Use a Flip-Flop as a basic memory cell.	3. Make use of temporary storage of data on the parasitic capacitances within the circuit.
4. Regenerating or re-circulating the data is not required.	4. Due to the leakage associated with junctions within the circuit, the charges on these capacitances leak off in a few μ sec. To prevent loss of data the charge must be restored or refreshed by regenerating or re-circulating the data. Refreshing of all bits is required almost every 2 mins.

(Contd...)

Static RAM	Dynamic RAM
5. Ordinary ICs are used.	5. Externally generated clock voltages required are too high to be generated using ordinary ICs. The separate ICs or discrete components are often used.
6. These are easy to interface to standard bipolar logic families.	6. Difficult to interface to standard bipolar logic families.
7. The chip area required per unit function is larger than that for dynamic RAMs	7. The chip area required per unit function is much smaller than that for static RAMs.
8. Static MOS circuit is less economical.	8. Dynamic MOS circuit is most economical.
9. Permits the integration of less amount of memory on a single chip.	9. Permits integration of large amount of memory on a single chip.
10. Power dissipation per function is greater than dynamic RAMs	10. Power dissipation per function is the lowest.

14

D/A Conversion and A/D Conversion

FACTS THAT MATTER

1. **D/A Conversion:** The process of converting a number of digital input signals to one equivalent analog output voltage.

2. **D/A converters:** Digital to analog (D/A) converter consists of resistive network, register level amplifier and some gating arrangement at the input of the register. The actual translation from digital signal to an analog signal takes place in the resistive network. D/A converter is considered to be a decoding device since it decodes the signals for entry into an analog system.

3. **Binary equivalent weight:** The value assigned to each bit in a binary (digital) number expressed as a fraction of the total. The binary weight of LSB is $1/(2^n-1)$ where n is the number of bits. The remaining weights are found by multiplying by 2, 4, 8 and so on.

4. **D/A converter testing:** There are two simple tests
   (*i*) The steady-state accuracy test
   (*ii*) The monotonicity test.
   (*i*) The steady-state accuracy test is setting a known digital number in the input register, measuring the analog output with an accurate meter and comparing with the theoretical value.
   (*ii*) The monotonicity test means checking that the output voltage increases regularly as the input digital signal increases. This can be done by using a counter as the digital input signal and observing the analog output on an oscilloscope. For proper

monotonicity, the output waveform should be a perfect staircase waveform with all steps equally spaced and of exact same amplitude.

Missing steps, steps of different amplitude or steps in a downward fashion shows that D/A converter is malfunctioning.

5. **Accuracy of D/A converter:** Accuracy is a measure of how close the actual output voltage is to the theoretical output value. The accuracy of D/A converter is primarily a function of the accuracy of the precision resistors used in the resistive network and the precision of the reference supply voltage used.

6. **Resolution:** I is the smallest increment in voltage that can be discerned. Resolution is primarily a function of the number of bits in the digital input signal or the smallest increment in output voltage is determined by the LSB.

7. **Millman's theorem:** It states that the voltage at any node in a resistive network is equal to the sum of the currents entering the node divided by the sum of the conductances connected to the node, all determined by assuming that the voltage at the node is zero.

In equation form

$$V_A = \frac{V_0/R_0 + V_1/R_1 + V_2/R_2 + V_3/R_3 + \cdots}{1/R_0 + 1/R_1 + 1/R_2 + 1/R_3 + \cdots}$$

8. **Binary ladder:** The binary ladder is a resistive network whose output voltage is a properly weighted sum of the digital inputs.

9. **A/D conversion:** The process of converting an analog input voltage to a number of equivalent digital output levels is known as analog to digital (A/D) conversion.

10. **A/D converter flash type:** This type of A/D converter effects fast and simultaneous conversion of analog data to digital through number of comparators.

11. **A/D converter tracking type:** This type of A/D converter effects tracking of analog input through its continuous comparison with converter's digital output.

12. **Quantization errors:** The smallest digital step or quantum is due to the LSB and can be made smaller only by increasing the number of bits in the counter. This inherent error is called the quantization error and is commonly ±1 bit.

13. **Differential linearity:** It is a measure of the variation in size of the input voltage to an A/D converter which causes the converter to change from one state to the next.

14. **Sample and hold circuit:** This circuit samples analog voltage signal and holds briefly to facilitate analog to digital conversion.

15. **SAR:** Successive approximation register consists of control logic and clock, ring counter, counter, level amplifiers. SAR is constructed on a single MSI chip. This chip is used in successive approximation A/D converter.

OBJECTIVE TYPE QUESTIONS

1. What is the LSB weight of a 5 bit resistive ladder?
2. What is a monotonicity test?
3. Why is a simultaneous A/D converter called a flash converter?
4. How many comparators will be required to build 5 bit simultaneous A/D converter?
5. In a D/A converter, LSB represents 0.2 V. The digital input 0111_2 will cause an analog output of _____ V.
6. What is the maximum and average conversion time for a counter type A/D converter driven by 1 MHz clock?
7. How does the continuous type A/D converter differ from the simple counter type A/D converter.
8. What advantage does the continuous type A/D converter offer over the counter-type A/D converter?
9. What does SAR stand for?
10. If multiplexing is required which A/D converter is most useful?
11. Is a single-ramp A/D converter slower or faster than a successive approximation A/D converter?
12. What advantage does the dual-slope A/D converter offer over the single ramp A/D converter?
13. What is an ADC0804?

Answers

1. $\dfrac{1}{2^5-1} = \dfrac{1}{31}$
2. A monotonicity test checks to see that the D/A output voltage increases regularly as the input digital signals increase.
3. Because its conversion time is very fast.
4. $2^5-1 = 31$ comparators will be required.
5. 1.4 V.
6. Maximum conversion time $= 1024 \times 10^{-6} = 1.024$ ms
   Average conversion time $= 0.512$ ms
7. The continuous type A/D converter uses an up-down counter.

8. The continuous type A/D converter is faster than the counter type A/D converter.

9. Successive approximation register.

10. Successive approximation converter (A/D).

11. Slower than successive approximation A/D converter.

12. The RC time constant cancels out making the conversion free from the absolute values of either R or C and also from variations in either value.

13. The ADC0804 is an 8 bit CMOS successive approximation A/D converter.

SHORT ANSWER TYPE QUESTIONS

Q.1. Find the binary equivalent weight of each bit in a 4-bit system.

Ans. The binary equivalent weight is the value assigned to each bit in a binary (digital) number expressed as a fraction of the total. The binary weight of LSB is $1/(2^n-1)$ where n is the number of bits. The remaining weights are found by multiplying by 2, 4, 8 and so on.

Hence for a 4 bit system, the LSB has a binary equivalent weight of $1/(2^4-1) = \dfrac{1}{15}$ or 1 part in 15. The 2nd LSB has a weight of $2 \times \dfrac{1}{15} = \dfrac{2}{15}$. The third LSB has a weight of $4 \times \dfrac{1}{15} = \dfrac{4}{15}$ and the MSB has a weight of $8 \times \dfrac{1}{15} = \dfrac{8}{15}$. The sum of the weights must be 1. Thus $\dfrac{1}{15} + \dfrac{2}{15} + \dfrac{4}{15} + \dfrac{8}{15} = 1$.

Q.2. What are the drawbacks of a resister divides network?

Ans. The resistive divider has two major drawbacks.

(*i*) Since each resistor in the network has a different value and the dividers are usually constructed using precision resistors, these divider network becomes very expensive.

(*ii*) The resistor used for the MSB is required to handle much greater current than that used for the LSB. For e.g. in a 10-bit system, the current through the MSB resistor is about 500 times as large as the current through the LSB. Such large current may damage the circuit.

LONG ANSWER TYPE QUESTIONS

Q.1. Explain the working of a resistive divider which may be used in D/A converter?

Ans. The basic requirement in converting a digital signal into an equivalent analog signal is to change the n digital voltage levels into one equivalent analog voltage. This can easily be achieved by designing a resistive divider that will change each digital level into an equivalent binary weighted voltage. Let us consider a resistive divider for a 3 bit system as shown in figure below.

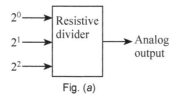

Fig. (a)

Let us assume that the digital input levels are 0 = 0 V and 1 = +7 V. Also let us consider the truth table for the 3 bit binary signals as shown in Fig. (b) below:

Now we want to change the eight possible digital signals in Fig. (b) into equivalent analog voltages. The smallest number here is 000.Let it be equal to 0v. the largest number is 111. Let this be equal to +7 V. so we now get the range of the analog signal to be from 0v to +7 V.

The binary equivalent weight of the LSB = $1/2^3 - 1 =$ $\dfrac{1}{8-1} = \dfrac{1}{7}$.

2^2	2^1	2^0
0	0	0
0	0	1
0	1	0
0	1	1
1	0	0
1	0	1
1	1	0
1	1	1

Fig. (b)

Thus the resistive divider will be designed such that a 1 in the 2° position will cause $+7 \times \dfrac{1}{7} = +1$ V at the output.

The smallest incremental change in the digital signal is represented by the LSB (2°).

Now since $2^1 = 2$ and $2^0 = 1$, clearly 2^1bit represents a number that is twice the size of the 2° bit. Hence a 1 in the 2^1 bit position will cause a change of $+7 \times \dfrac{2}{7} = +2$ V in the analog voltage. Similarly 1 in the 2^2 bit position will cause a change of $+7 \times \dfrac{4}{7} = +4$ V.

The digital input 011 is the combination of the signals 001 and 010. So if we add +1 V from 2^0 bit and +2 V from the 2^1 bit, the output of +3 V for input 011 is achieved. The other voltage levels for the respective input digital levels are shown in Fig. (c) below:

A resistive divider thus must do two things in order to change the digital input into an equivalent analog output voltage.

Digital input			Analog output
0	0	0	+0 V
0	0	1	+1 V
0	1	0	+2 V
0	1	1	+3 V
1	0	0	+4 V
1	0	1	+5 V
1	1	0	+6 V
1	1	1	+7 V

Fig. (c)

(*i*) The 2^0 bit must be changed to +1 V, 2^1 bit to +2 V and 2^2 bit to +4 V.

(*ii*) These three voltages representing the digital bits must be summed together to form the analog output voltage.

Such a resistive divider is shown below in Fig. (*d*).

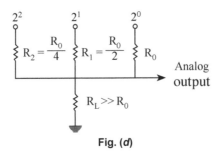

Fig. (*d*)

Resistors R_0, R_1 and R_2 form the divider network. Resistance R_L represents the load to which the divider is connected which is large enough so it does not load the divider network.

Now let us assume that digital signal 001 is applied to the divider network. Since 0 = 0 V and 1 = +7V (as we have assumed), the equivalent circuit becomes as follows:

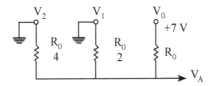

$R_L \gg R_0$ so it is neglected. The analog output voltage V_A can be easily found out by using Millman's theorem, which states that the voltage appearing at any node in a resistive network is equal to the summation of the currents entering the node (assuming the node voltage is zero) divided by the summation of the conductances connected to the node *i.e.*

$$V_A = \frac{V_0/R_0 + V_1/(R_0/2) + V_2/(R_0/4)}{1/R_0 + 1/(R_0/2) + 1/(R_0/4)}$$

$$= \frac{7/R_0 + 0 + 0}{1/R_0 + 2/R_0 + 4/R_0} = \frac{7/R_0}{7/R_0} = +1V$$

Similarly if we draw the equivalent circuits for the other 7 input combinations and apply Millman's theorem, we will get the corresponding voltages as shown in the table in Fig. (*c*) above. The change in output voltage due to a change in the LSB is equal to $V/(2^n-1)$, where V is the digital input voltage level and *n* is the number of bits.

Q.2. What is a binary ladder? Explain the working of a binary ladder designed for 4 bits.

Ans. The binary ladder is a resistive network whose output voltage is a properly weighted sum of the digital inputs. Such a ladder designed for 4 bits is shown below:

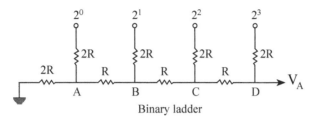

Binary ladder

We find that binary ladder is constructed which has only two values of resistors. Thus it overcomes one of the drawback of resistive divider network.

The left end of the ladder is terminated in a resistance of 2R and let us assume that the right end *i.e.* the output is open circuited.

Let us first find out the resistive properties of the network assuming all the digital inputs are at ground. If we stand at node A, the total resistance looking into the terminating resistor is 2R and the total resistance looking out toward the 2^0 input is also 2R. These two resistors can be combined to form an equivalent resistor of value R. Likewise if we stand at any node, the total resistance looking back toward the terminating resistor or out toward the digital input is 2R. This is true regardless of whether the digital inputs are at ground or +V, because the internal impedance of an ideal voltage source is 0Ω and we are assuming that the digital inputs are ideal voltage sources.

By applying the resistance characteristics of the ladder we can determine the output voltages for the various digital inputs. Let us first take the digital input as 1000. With this input the binary ladder can be redrawn as shown below in Fig. (*a*).

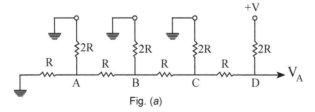

Fig. (*a*)

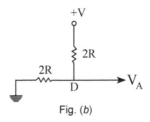

Fig. (b)

Since there are no voltage sources to the left of node D, the entire network to the left of this node can be replaced by a resistance of 2R.

Thus the equivalent circuit is shown in Fig. (b) above.

From this equivalent circuit, it can be seen that the output voltage is

$$V_A = V \times \frac{2R}{2R + 2R} = +\frac{V}{2}$$

Thus an 1 in the MSB position will provide an output voltage of $+\dfrac{V}{2}$.

Now to determine the output voltage due to the second MSB, let us assume a digital input 0100. The corresponding circuit is shown below in Fig. (c).

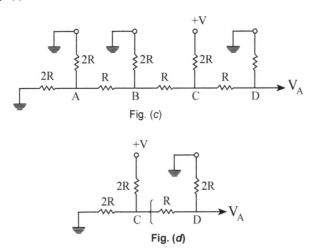

Fig. (c)

Fig. (d)

Since there are no voltage sources to the left of node C, the entire network to the left of node C can be replaced by 2R as shown in Fig. (d) above. Let us now replace the network to the left of node C with its Thevenis equivalent by cutting the circuit on the jagged line as shown in Fig. (d). The Thevenin equivalent is simply a resistance R in series with a voltage source of +V/2. The final equivalent circuit is shown in Fig. (e) below.

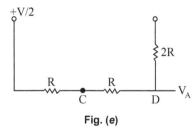

Fig. (e)

So, from Fig. (e) we find $V_A = \dfrac{+V}{2} \times \dfrac{2R}{R+R+2R} = +\dfrac{V}{4}$. Thus the second

MSB provides an output voltage of $+\dfrac{V}{4}$.

If we continue in this manner, we will find that third MSB provides an

output of $+\dfrac{V}{8}$, the fourth MSB provides an output voltage of $+\dfrac{V}{16}$ and

so on. The output voltages for the binary ladder are shown in the table

below:

Bit position	Binary weight	Output voltage
MSB	1/2	V/2
2nd MSB	1/4	V/4
3rd MSB	1/8	V/8
4th MSB	1/16	V/16
5th MSB	1/32	V/32
⋮	⋮	⋮
nth MSB	$1/2^n$	$V/2^n$

Since this ladder is composed of linear resistors, it is a linear network
and principle of superposition can be used. Therefore the total output
voltage will be equal to the sum of the output voltages caused by each
digital input individually.

In equation form the output voltage is given by

$$V_A = \frac{V}{2} + \frac{V}{4} + \frac{V}{8} + \frac{V}{16} + \cdots + \frac{V}{2^n} \qquad \text{...(i)}$$

where n is the total number of bits at the input.

The above equation can be simplified and the simplified equation for the
output voltage can be given in the form

$$V_A = \frac{V_0 2^0 + V_1 2^1 + V_2 2^2 + \cdots + V_{n-1} 2^{n-1}}{2^n} \qquad \text{...(ii)}$$

where $V_0, V_1, V_2, \ldots V_{n-1}$ are the digital input voltage levels and n is the number of bits. This equation (ii) can be used to find the output voltage from the
ladder for any digital input signal.

Q.3. Explain the working of a 4 bit D/A converter.

Ans. The basis for a digital to analog (D/A) converter is the use of a resistive divider or a binary ladder. It is in the resistive network that the actual translation from a digital signal to an analog voltage takes place. With some additional circuitry we can complete the design of D/A converter. The block diagram of the complete D/A converter is shown below in Fig. (*a*).

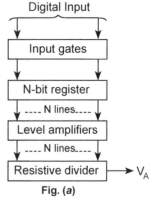

Fig. (*a*)

There must be a register as an integral part of the D/A converter. This register can be used to store the digital information. This register can be of any one of the many types. Simplest register is formed by use of RS flip-flops with one flip-flop per bit. There must be some gating arrangements at the input of the register such that the flip-flops can be set with proper information from the digital system. Next there must be level amplifiers between the register and the resistive network to ensure that the digital signals presented to the network are all of the same level. The complete schematic diagram for a 4 bit D/A converter is shown below in Fig. (*b*).

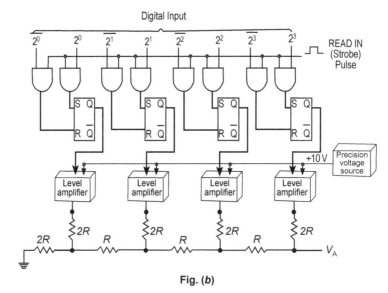

Fig. (*b*)

From the fig it is seen that the four flip-flops form the register necessary for storing the 4 bit digital information. The flip-flop on the right represents the MSB and the flip-flop on the left represents the LSB. Each flip-flop is a simple RS latch and requires a positive level at the R or S input to reset or set it. The gating arrangement for entering information into the register is such that it meets the requirement. Also with this gating scheme, the flip flops need not be reset or set each time new information is entered. When the READ IN line goes high, only one of the two gate outputs connected to each flip-flop is high and the flip-flop is set or reset accordingly. Thus the data are entered into the register each time the READ IN or strobe pulse occurs.

Now the output from the flip-flops goes to the input of respective level amplifier. The other input of the level amplifier is the +10 V from the precision voltage source. The amplifiers work in such a way that when the input from a flip-flop is high, the output of the amplifier is at +10 V. When the input from the flip-flop is low, the output is OV.

Finally the output of the level amplifiers is fed to the resistive network. In other words the digital input is fed to the resistive network. In the fig above this resistive network is taken as a binary ladder. The binary ladder converts the digital input into equivalent analog voltage and we get the analog output using the equation.

$$V_A = \frac{V_0 2^0 + V_1 2^1 + V_2 2^2 + V_3 2^3}{2^4}$$

Thus if the digital input was 1011, we will get the analog output voltage

$$V_A = \frac{10 \times 2^0 + 10 \times 2^1 + 0 \times 2^2 + 10 \times 2^3}{2^4}$$

$$(\because 1 = +10V \text{ and } 0 = 0V)$$

$$= \frac{10 \times 1 + 10 \times 2 + 0 + 10 \times 8}{16}$$

$$= \frac{10 + 20 + 80}{16} = \frac{110}{16}$$

$$= +6.875 \text{ V}$$

Q.4. Explain the working of 2 bit simultaneous A/D converter.

Ans. The process of converting an analog voltage into an equivalent digital signal is known as analog to digital (A/D) conversion.

There are a number of methods available to do the same, the simplest of which is the simultaneous method.

The simultaneous method of A/D conversion is based on the use of a number of comparator circuits. The number of comparators required is determined by $2^n - 1$ where n is the number of bits. Therefore for 2 bit simultaneous A/D converter, number of comparators required will be $2^2 - 1 = 4 - 1 = 3$ comparators. The block diagram for 2 bit simultaneous A/D converter is shown below:

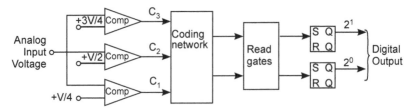

From the block diagram are we find that the analog signal to be digitized is one of the inputs to each comparator and the second input is a standard reference voltage. The reference voltages used here are $+\dfrac{V}{4}$, $+\dfrac{V}{2}$ and $+3\dfrac{V}{4}$ such that the system is capable of accepting an analog input voltage which ranges from to 0 to +V volt.

The comparator works in such a way that if the analog input signal exceeds the reference voltage to any comparator, the output of that comparator is high. Now, if all the comparators' outputs are low, the analog input must be between 0 and $+\dfrac{V}{4}$. If C_1 is high and C_2 and C_3 are low the input must be between $+\dfrac{V}{4}$ and $+\dfrac{V}{2}$. If C_1 and C_2 are high and C_3 low, the input must be between $+\dfrac{V}{2}$ and $+3\dfrac{V}{4}$. If all the comparator outputs are high, the input voltage must be between $+3\dfrac{V}{4}$ and +V. The corresponding table is shown below:

Input voltage	Comparator output		
	C_1	C_2	C_3
0 to $+\dfrac{V}{4}$	0	0	0
$+\dfrac{V}{4}$ to $+\dfrac{V}{2}$	1	0	0
$+\dfrac{V}{2}$ to $+3\dfrac{V}{4}$	1	1	0
$+3\dfrac{V}{4}$ to +V	1	1	1

Thus these four voltage ranges can be detected by this converter. The four ranges can be discerned by two binary bits. The three comparator outputs are first fed to the coding network. The output of the coding network are 2 bits which are equivalent to the input analog voltage. These 2 bits are sent to the gating arrangement, and the proper READ gate goes high and the corresponding 2 bits enter the flip-flop register for storage. In this case the 2 bits stored will be any one of the following at a time: 00, 01, 10 or 11 according to the input voltage.

Q.5. With the help of logic diagram explain the working of 3 bit simultaneous A/D converter.

Ans. The process of converting an analog voltage into an equivalent digital signal is known as analog to digital (A/D) conversion.

There are many methods to do the same. The simplest method is the simultaneous method.

The simultaneous method of A/D conversion is based on the use of number of comparator circuits. The number of comparators required is equal to 2^n-1 where n is the number of bits. Thus 3 bit simultaneous A/D converter will require $2^3-1 = 8 - 1 = 7$ comparators. The logic diagram for 3 bit simultaneous A/D converter is shown below:

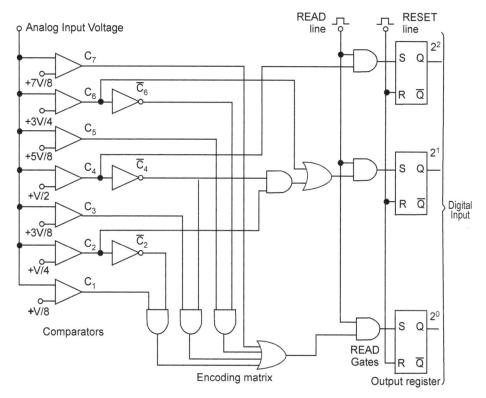

From the logic diagram we see that some of the comparators have inverters at their outputs since both C and $\overline{C}$ are needed for the encoding matrix.

The encoding matrix must accept seven input levels and encode them into a 3 bit binary number having eight possible states. The operation of the encoding matrix can be understood by the table given below:

Input voltage	Comparator output							Binary output		
	C_1	C_2	C_3	C_4	C_5	C_6	C_7	2^2	2^1	2^0
0 to $+\frac{V}{8}$	0	0	0	0	0	0	0	0	0	0
$\frac{V}{8}$ to $\frac{V}{4}$	1	0	0	0	0	0	0	0	0	1
$\frac{V}{4}$ to $\frac{3V}{8}$	1	1	0	0	0	0	0	0	1	0
$\frac{3V}{8}$ to $\frac{V}{2}$	1	1	1	0	0	0	0	0	1	1
$\frac{V}{2}$ to $\frac{5V}{8}$	1	1	1	1	0	0	0	1	0	0
$\frac{5V}{8}$ to $\frac{3V}{4}$	1	1	1	1	1	0	0	1	0	1
$\frac{3V}{4}$ to $\frac{7V}{8}$	1	1	1	1	1	1	0	1	1	0
$\frac{7V}{8}$ to V	1	1	1	1	1	1	1	1	1	1

Logic table for the 3 bit converter.

There are two inputs to each comparator. One input is the analog signal to be digitized and the other is a standard reference voltage. The reference voltages are V/8, V/4, 3V/8, V/2, 5V/8, 3V/4 and 7V/8 such that the system is capable of accepting an analog input voltage which range from 0 to +V volt. The comparator works in such a way that if the analog input signal exceeds the reference voltage to any comparator, the output of that comparator is high or 1. Now if all the comparator outputs are low or 0, the analog input must be between 0 and +V/8. If C_1 is high and all other comparator outputs are low, the input must be between +V/8 and +V/4. If C_1 and C_2 are high and all other five comparator outputs are low, the input voltage must be between +V/4 and +3V/8. If we proceed in a similar way, when all the comparator outputs are high, the input voltage must be between +7V/8 to +V volt.

The comparator outputs reach the 3 bit binary register consisting of R S flip-flops via the encoding matrix. The values of the register outputs 2^2, 2^1 and 2^0 depends on the input voltage level.

The 2^2 bit is high or the 2^2 flip-flop is set whenever C_4 is high.

The 2^1 time line will be high whenever C_2 is high and $\overline{C}_4$ is high or whenever C_6 is high. In equation form $2^1 = C_2\overline{C}_4 + C_6$.

Similarly the logic equation for 2^0 is

$$2^0 = C_1\overline{C}_2 + C_3\overline{C}_4 + C_5\overline{C}_6 + C_7$$

To transfer the data from the encoding matrix to the register is carried out in two steps as follows:

(i) A positive reset pulse must appear on the RESET line to reset all the flip-flops low.

(ii) Next a positive READ pulse allows the proper READ gates to go high or low and accordingly transfer the digital information into the flip-flops.

The simultaneous method is simple and conversion rate is extremely fast. Because it is so fast, this type of converter is also known as flash converter.

The drawback of simultaneous method is that as the number of bits in the desired digital number increases, the number of comparators required increases very rapidly ($2^n - 1$) and so it becomes unmanageable.

Q.6. Explain the working of counter type A/D converter.

Ans. Counter type A/D converter uses only one comparator and a variable reference voltage. This reference voltage is applied to the comparator and when it becomes equal to the input analog voltage, the conversion is complete. The block diagram of the converter is shown below:

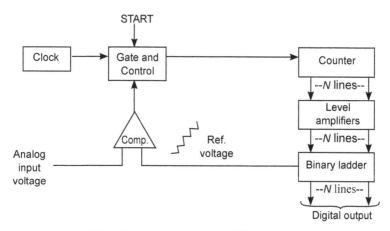

Block diagram of counter type A/D converter

If we notice the block diagram carefully, we see that this converter consists of a D/A converter (the counter, level amplifiers and the binary ladder), one comparator, a clock and the gate and control circuitry.

The digital output signals will be taken from the counters and it should be an n bit counter where n is the desired number of bits. The operation of the converter is a follows. First, the counter is reset to all 0s. Now when a convert signal appears on the START line, the gate opens and the clock pulses are allowed to pass through to the input of the counter. The counter advances through its normal binary count sequence and at the output of the binary ladder we get the staircase waveform. This waveform acts as the reference voltage which is fed to the comparator input. The other input of the comparator is the analog input voltage. When the reference voltage equals or exceeds the input analog voltage, the gate is closed which is controlled by the control circuitry, the counter stops and the conversion is complete. The number stored in the counter is now the digital equivalent of the analog input voltage.

An error signal is generated at the output of the comparator by taking the difference between the analog input signal and the reference staircase voltage. The error is detected by the control circuit and the clock is allowed to advance the counter. As the counter advances the error signal is reduced by increasing the reference voltage. When the error is reduced to zero, the reference voltage is equal to the analog input signal, the control circuitry stops the clock from advancing the counter.

The counter type A/D converter is a very good method for digitizing to a high resolution but the conversion time required is much longer than simultaneous A/D converter.

Since the counter always begins at zero and counts through its normal sequence, as many as 2^n counts may be necessary before conversion is complete (n is the required no of bits). For example if we have a 10 bit converter, it requires 1024 clock cycles for a full scale count. If we use 1MHz clock, full scale count requires $1024 \times 10^{-6} = 1.024$ms.

Q.7. Explain the working of continuous A/D converter. What advantage does the continuous type A/D converter offer over the counter type A/D converter.

Ans. In the counter type A/D converter the resolution is very high but the conversion time is much longer. For speeding up the conversion time, the need for resetting the counter each time a conversion is made, should be eliminated. To achieve this, we should use up-down counter which is capable of either counting up or down such that resetting is not required. Along with the up- down counter we need additional circuitry for A/D converter as shown in fig. below. This method is known as continuous conversion since resetting is not required and thus the converter is called a continuous type A/D converter. Since the converter's digital output always tries to track the analog input, this is also known as A/D converter tracking type.

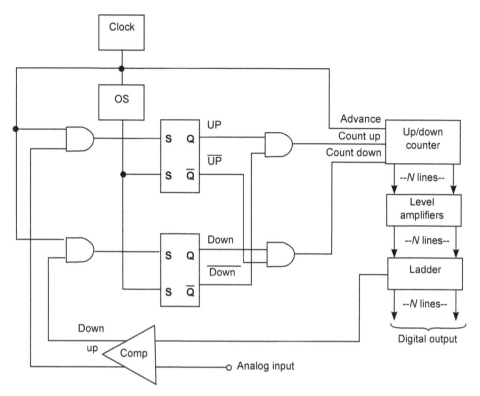

From the figure above we find that the D/A portion used in this converter consists of up-down counter, level amplifiers and binary ladder. The output of the ladder is fed to a comparator which has two inputs and two outputs. The other input fed is the analog voltage. When the analog voltage is more positive than the ladder output, the up output of the comparator is high and when the analog voltage is more negative than the ladder output, the down output is high.

Now the AND gate which controls the count-up line has inputs of up and $\overline{down}$ and the count-down line And gate has inputs of down and $\overline{up}$. This is actually an exclusive OR arrangement which ensures that count-up line and count-down line cannot both be high at the some time.

Now if the up output of the comparator is high, the upper AND gate whose output is fed to the Up flip-flop is enabled and the first time the clock goes positive, the up flip-flop is set. The true output of the flip-flop is fed to the count-up line AND gate whose output *i.e.,* count up line will be high and the counter will advance one count. The counter can advance only one count since the output of the one-shot (OS) resets both the up and down flip-flops just after the clock goes low. This is one count-up conversion cycle.

As long as the up line out of the comparator is high, the converter continues to operate one conversion cycle at a time. At the point when the analog voltage becomes more negative than the ladder voltage, the up line of the comparator goes low and down line goes high. The converter then goes through a count-down conversion cycle. At this, the ladder voltage is within 1 LSB of the analog voltage, and the converter oscillates about this point.

To avoid the oscillation, the comparator is adjusted by centering on the LSB. This means that the up output will not go high unless the ladder voltage is more than ½ LSB below the analog voltage. Similarly the down output will not go high unless the ladder voltage is more than ½ LSB above the analog voltage.

Thus we can conclude that this converter can convert the analog input voltage into equivalent digital voltage continuously without the need of resetting the counter. Thus the advantage of the continuous type A/D converter is that it is much faster than the counter type A/D counter in which resetting the counter is needed.

Q.8. If multiplexing is required, which method is used for A/D conversion. Explain its working.

Ans. The continuous A/D converter is very fast once it is locked on the signal but loses this advantage if multiplexing inputs are required.

If multiplexing is required, the successive approximation converter is most useful. The block diagram for the converter is shown below in Fig (*a*) and operation is shown in Fig (*b*).

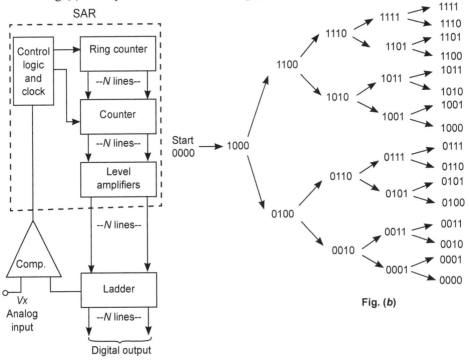

Fig. (*a*)

Fig. (*b*)

The converter consists of SAR, Ladder and comparator SAR consists of ring counter, counter, level amplifiers, clock and control logic. SAR is successive approximation register.

The converter operates by successively dividing the voltages ranges in half.

The inputs of the comparator are the analog voltage and the output of the binary ladder. The counter is first reset to all 0s and the MSB is then set. The MSB is then left in or taken out by resetting the MSB flip-flop with the next clock, depending on the output of the comparator. Then the second MSB is set in and a comparison is made to determine whether to reset the second MSB flip-flop or not. The process is repeated till we reach the LSB and at this time the desired number is in the counter. Since the conversion involves operating on one flip-flop at a time, beginning with the MSB a ring counter is used for flip-flop selection.

Thus the successive approximation method is the process of approximating the analog voltage by trying 1 bit at a time beginning with the MSB. The operation can easily be understood by observing Fig. (*b*) above. Here each conversion taken the some time and requires one conversion cycle for each bit. Thus the total conversion time is equal to *n* times the time required for one conversion (*n* being the number of bits). One conversion cycle normally requires one cycle of the clock. So if we use 1 MHz clock the 10 bit converter will have a conversion time equal to $10 \times 10^{-6} = 10$ μs. Thus this method of conversion is very fast.

SAR is generally available as a single MSI chip. Motorola MC 6108 is an example of such a chip.

Q.9. Explain the working of single slope A/D converter.

Ans. If a very short conversion time is not a requirement, single-slope A/D conversion method is very economical and simple. This type of converter does not use D/A converters and makes use of a linear ramp generator to produce a constant slope reference voltage. This method involves comparison of the unknown input voltage with the reference voltage that begins at zero and increases linearly with time. The time required for the reference voltage to increase to the value of the unknown voltage is directly proportional to the magnitude of the unknown voltage and this time period is measured with a digital counter.

The most important part of this method is the ramp generator. The ramp generator can be constructed using an operational amplifier (OA) connected as an integrator as shown below:

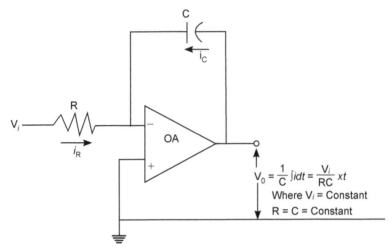

$$V_0 = \frac{1}{C} \int i\,dt = \frac{V_i}{RC} \times t$$

Where V_i = Constant

$R = C$ = Constant

Since V_i, R and C are constants, this equation is of a straight line that has a slope $\dfrac{Vi}{RC}$.

The logic diagram and wave form of the single slope converter are shown below in Fig. (*a*) and (*b*).

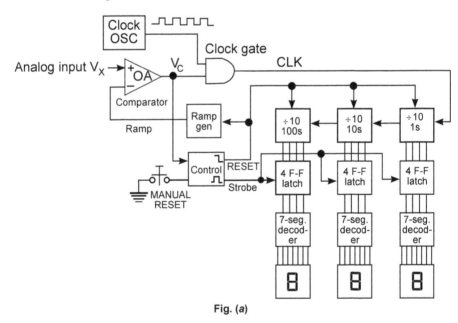

Fig. (*a*)

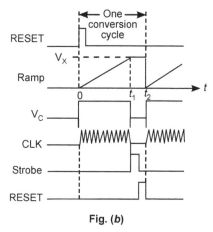

Fig. (b)

Now let us see how this converter works. Let us assume that the clock is running continuously and that the input voltage V_X we wish to digitize is positive. The three decade counters are connected in cascade and their outputs can be strobed into three 4 flip-flop latch circuits. The latches are then decoded by seven segment decoders which drive the LED displays as units, tens and hundreds of counts.

The conversion cycle begins by first depressing the MANUAL RESET switch. This generates a reset pulse which resets all the decade counters to zeros and the ramp voltage to zero. Since V_X is positive and RAMP begins at zero, the output of the comparator OA, V_C must be high. This voltage enables the clock gate allowing the CLK to be applied to the decade counter. The counter begins counting upward and the RAMP continues upward until ramp voltage is equal to the unknown voltage V_x. As we see from the waveforms, at this point t_1 (time), the comparator output V_C goes low, thus disabling the CLOCK gate and the counter stops. Simultaneously this negative transition on V_C generates a STROBE signal in the CONTROL box that shifts the contents of the decade counters into the 4 flip-flop latches. The contents of the latches are decoded by 7-segment decoders and are displayed on 7-segment LEDS. Shortly thereafter a reset pulse is generated by the CONTROL box that reset the RAMP and clears the decade counters to 0s and another conversion cycle can began.

This converter as it stands can display the value of input voltage between 0 to 999 mv.

Since the reference voltage in this method is sloped like a ramp, this is referred to a single ramp method also.

One drawback of this single slope method is its dependency on an extremely accurate ramp voltage which in turn is dependent on values of R and C. The values of R and C may vary with time and temperature.

Q.10. Explain the working of Dual slope A/D converter.

Ans. The Dual slope A/D converter overcomes the problem of accuracy of single slope A/D converter. Single slope A/D converter is strongly dependent on the values of R and C, but Dual-slope A/D converter does not depend on R&C and so, this is the most accurate method. The logic diagram of Dual slope A/D converter is shown below

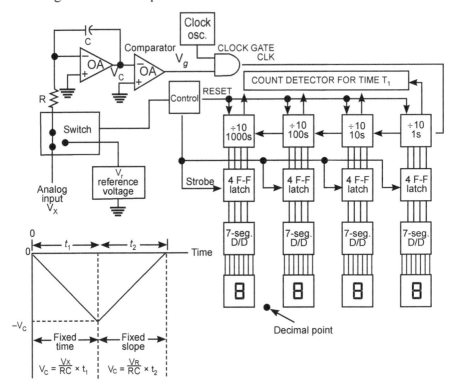

From the logic diagram we see that the operational amplifier connected as integrator forms the desired ramp. This integrator forms two different ramps—first when the input switch is connected to the unknown input voltage V_x and second when it is connected to a known reference voltage V_r. The other OA acts as the comparator.

Let us now see how this converter works: Let us assume that the clock is running continuously and input voltage V_x is positive. First the decade counters are cleared to all zeros and the ramp reset to 0.0 V. Now the input is switched to the unknown input voltage V_x. Since V_x is positive, from the diagram we find that the integrator output V_C is negative ramp. The comparator output V_g is thus positive and so clock is allowed is pass through the clock gate to the decade counters. Let the ramp be allowed to proceed for a fixed time t_1. So V_C at the end of fixed time t_1 will be equal to $-\dfrac{V_X}{RC} \times t_1$ (output of the integrator).

When the counter reaches the fixed count at time t_1, the CONTROL unit generates a pulse to clear the decade counters to all 0s. The integrator input is now switched to the negative reference voltage V_r. The integrator will now begin to generate a ramp starting at $-V_C$ and increasing steadily upward until it reaches 0.0 V. All this time the counter is counting and conversion cycle ends when $V_C = 0.0$ V since the clock gate get disabled now. The equation of the positive ramp is $V_C = \dfrac{V_r}{RC} \times t_2$. Here $\dfrac{V_r}{RC}$ i.e. slope of ramp is constant but t_2 is variable.

Thus we find that the integrator output voltage begins at 0.0v, goes down to $-V_C$ and then again goes back to 0.0v. So we can equate the two equations for V_C i.e.

$$\frac{V_r}{RC} \times t_1 = \frac{V_r}{RC} \times t_2$$

$$\text{or} \quad V_x = V_r \times \frac{t_2}{t_1}$$

Since V_r and t_1 is known or constant the unknown voltage is directly proportional to variable time period t_2. This time period t_2 is exactly the contents of the decade counters at the end of a conversion cycle.

To understand the operation properly, let us take an example:

Let the clock be 1.0 MHz, the reference voltage $V_r = -1.0$ V dc, $t_1 = 1000$ μs, $R_C = 1.0$ ms. During time t_1, $V_C = -1.0$ V dc if $V_X = 1.0$ V. Then during time t_2, V_C will ramp all the way back to 0.0 V and this will require a time period of $t_2 = 1000$ μs, since the slope of this ramp is fixed at 1.0 V/ms.

The output display will now read 1000. Actually as soon as the clock gate gets disabled the counter stops. Simultaneously the CONTROL unit generates a STROBE signal that shifts the contents of the decade counters into the 4 flip-flop latches. The contents of the latches are decoded by 7 segment decoders and are displayed on 7 segment LEDs. Shortly thereafter a reset pulse is generated by the CONTROL unit that resets the decade counters to 0s and another conversion cycle can start.

Now in the above example the display we got was 1000. With placement of decimal point as shown in logic diagram, this will read 1.000 V.

Thus we find that Dual slope A/D converter is free from R and C and thus this is much more accurate than single slope A/D converter.